Rehab Mansour

Sorption of Xenobiotics to Humic Acid in Aqueous Systems

Rehab Mansour

Sorption of Xenobiotics to Humic Acid in Aqueous Systems

Südwestdeutscher Verlag für Hochschulschriften

Impressum / Imprint

Bibliografische Information der Deutschen Nationalbibliothek: Die Deutsche Nationalbibliothek verzeichnet diese Publikation in der Deutschen Nationalbibliografie; detaillierte bibliografische Daten sind im Internet über http://dnb.d-nb.de abrufbar.

Alle in diesem Buch genannten Marken und Produktnamen unterliegen warenzeichen-, marken- oder patentrechtlichem Schutz bzw. sind Warenzeichen oder eingetragene Warenzeichen der jeweiligen Inhaber. Die Wiedergabe von Marken, Produktnamen, Gebrauchsnamen, Handelsnamen, Warenbezeichnungen u.s.w. in diesem Werk berechtigt auch ohne besondere Kennzeichnung nicht zu der Annahme, dass solche Namen im Sinne der Warenzeichen- und Markenschutzgesetzgebung als frei zu betrachten wären und daher von jedermann benutzt werden dürften.

Bibliographic information published by the Deutsche Nationalbibliothek: The Deutsche Nationalbibliothek lists this publication in the Deutsche Nationalbibliografie; detailed bibliographic data are available in the Internet at http://dnb.d-nb.de.
Any brand names and product names mentioned in this book are subject to trademark, brand or patent protection and are trademarks or registered trademarks of their respective holders. The use of brand names, product names, common names, trade names, product descriptions etc. even without a particular marking in this works is in no way to be construed to mean that such names may be regarded as unrestricted in respect of trademark and brand protection legislation and could thus be used by anyone.

Coverbild / Cover image: www.ingimage.com

Verlag / Publisher:
Südwestdeutscher Verlag für Hochschulschriften
ist ein Imprint der / is a trademark of
OmniScriptum GmbH & Co. KG
Heinrich-Böcking-Str. 6-8, 66121 Saarbrücken, Deutschland / Germany
Email: info@svh-verlag.de

Herstellung: siehe letzte Seite /
Printed at: see last page
ISBN: 978-3-8381-3942-5

Zugl. / Approved by: Freiberg, TU Bergakad., Diss., 2012

Copyright © 2014 OmniScriptum GmbH & Co. KG
Alle Rechte vorbehalten. / All rights reserved. Saarbrücken 2014

Preface

The current thesis results obtained during my PhD project carried out in the Department of Ecological Chemistry of the Helmholtz Centre for Environmental Research-UFZ, Leipzig, Germany during the time from April 2007 to March 2012. I have been enrolled as a PhD student at Technische Universität Bergakademie Freiberg.

My PhD. project has been financed by the Egyptian High Ministry of Education, Cultural Affairs and Missions Sector, as governmental PhD-scholarships program in collaborative with Ain Sham University, Cairo, Egypt.

The other part through my participation in the EU project OSIRIS "Optimized strategies for risk assessments of industrial chemicals through integration of non test and test information" which was coordinated by Prof. Dr. Gerrit Schüürmann at the UFZ-Department of Ecological Chemistry.

Acknowledgements

This work was funded by the the High Ministry of Education and Ain Shams University, Cairo, Egypt. The funding of this research is gratefully acknowledged. I am so glad to the Egyptian government for giving myself chance to study abroad to follow my scientific carrier and update the new knowledge and have the opportunity to come to the Ecological Chemistry group in Leipzig.

I would like to express my sincere gratitude and greatest thanks go to my supervisor and my promoter in Germany Prof. Dr. Gerrit Schüürmann for having supervised my PhD. Always approachable and encouraging, his honesty and reliability have been much appreciated and helped me maintain confidence and motivation throughout this project.

My deepest gratitude and thanks to Dr. Albrecht Paschke for his supervision, continuous support, helpful advices, and constructive criticism and encouragement in order to complete my thesis during my study in Leipzig.

I am deeply thanking and indebted to Mr. Uwe Schröter for his continues help in developing the analytical methods, solving the technical equipment problems and his moral support during my study in Leipzig.

I am grateful to Dr. Ralph Kühne, Mr. Ralf-Uwe Ebert and Mr. Dominik Wondrousch for their useful discussion, careful guidance in modelling and statistical analysis and for allowing access to the "UFZ in-house version of ChemProp".

I would like to thank also Dr. Wolfgang Geyer and Department of Analytical Chemistry for their help in FTIR and elemental analysis.

I would like express my sincere gratitude for all the PhD students and all group members during my study in the Department of Ecological Chemistry, Helmholtz Centre for Environmental Research-UFZ in Leipzig.

Finally, I would like to thank all my family their continuous encouragement, help and support.

List of Contents

Contents i
List of Abbreviations and Symbols iv
List of Tables vi
List of Figures and Schemes viii

1. **Introduction** 1
   1.1 Scope of the problem 1
   1.2 Sorption of xenobiotics to soils and sediments 2
   1.3 Sorption of humic acid 3
   1.4 Selection of probe compounds 4
   1.5 Literature review of sorption process 4
         1.5.1 Sorption of xenobiotics in the aquatic environment 5
         1.5.2 Influence of pH on sorption process 10
         1.5.3 Influence of temperature on sorption process 13
   1.6 The importance of the topic for the Egyptian environment in the future 15
   1.7 The importance of this topic for REACH in Europe 16
   1.8 Scope and outline of the thesis 16

2. **Theoretical Background** 18
   2.1 Scope in sorption phenomena 18
   2.2 Sorption isotherms 19
   2.3 Sorption mechanisms 22
   2.4 Factors affecting sorption coefficient 24
   2.5 Experimental and estimation methods for determination of sorption coefficient 26
         2.5.1 Experimental methods 26
         2.5.2 Estimation methods 28
   2.6 Nernst's Distribution Law 30
   2.7 Equilibrium partitioning of pollutants in the environment 31
   2.8 pH Effect on equilibrium partitioning 32
   2.9 Temperature effect on equilibrium partitioning 34
   2.10 Humic substances 36
   2.11 Humic substances role in soil photolysis 37

3. Materials and Methods 39

3.1 Materials 39
3.1.1 Organic chemicals 39
3.1.2 Equipments 40

3.2 Experimental methods 45
3.2.1 SPME optimization 45
3.2.2 Check for saturation effect (isotherm measurement) 49
3.2.3 Batch method for sorption experiments 50
3.2.3.1 Direct immersion mode SPME-GC-MSD (in full vial without headspace) 52
3.2.3.2 Headspace mode HS-SPME-GC-MSD (in half-filled vial) 53
3.2.4 GC-MSD analysis 53
3.2.5 Determination of K_{oc} 57

3.3 Comparison with mean log K_{oc} in literature 60

3.4 QSAR modeling validation and predictability 61
3.4.1 Log K_{ow}-log K_{oc} correlationship 61
3.4.2 Linear Solvation Energy Relationship (LSER) 62
3.4.3 Comparison of experimental and calculated log K_{oc} from different prediction methods 64
3.4.4 Statistics 65
3.4.4.1 Regression analysis 65
3.4.4.2 Performance parameters 66

4. Results and Discussion 67

4.1 A closer look at sorption of xenobiotics to humic acid in neutral water 67
4.1.1 Introduction 67
4.1.2 Comparison with mean log K_{oc} in literature 67
4.1.3 Comparison with calculated log K_{oc} from different prediction methods 74
4.1.5 Refitting of the common K_{ow}-approach for sorption to humic acid 86
4.1.6 Refitting of LSER-approach for sorption to humic acid 90
4.1.7 Conclusions 100

4.2 pH-Dependence of sorption **101**

4.2.1 Introduction 101
4.2.2 Comparison with mean log K_{oc} in literature 101
4.2.3 Comparison with calculated log K_{oc} from different prediction methods 108
4.2.4 Refitting of the common K_{ow}-approach for sorption to humic acid 109
4.2.5 The influence of pH variation on the values of log K_{oc} (effect of solute structure, physical properties on partitioning processes according to Henderson-Hasselbach equation) 115
4.2.6 Refitting of LSER-approach for sorption to humic acid 120
4.2.7 The suggested model to predict log K_{oc} based on pH of for neutral, acidic and basic compounds, and its comparison with the Franco and Trapp model 125
4.2.8 Conclusions 133

4.3 Temperature-dependence of sorption **134**

4.3.1 Introduction 134
4.3.2 Sorption and temperature effect 134
4.3.3 Thermodynamic parameters based on the mole fraction partition coefficients K_{oc}^x 135
4.3.4 Comparison with mean log K_{oc} in literature 143
4.3.5 Comparison with calculated log K_{oc} from different predictions methods 151
4.3.6 Refitting of the common K_{ow}-approach for sorption to humic acid 153
4.3.7 Refitting of LSER-approach for sorption to humic acid 155
4.3.8 Conclusions 159

5. Summary and Conclusions **160**
6. Zusammenfassung und Schlussfolgerungen **163**
7. Future Perspectives **167**
8. Citations and Bibliography **168**
9. Appendix **178**

List of Abbreviations and Symbols

Abbreviations common used in this thesis, listed in alphabetical order.

$\Delta G^x_{w \to humic}$	The free Gibbs energy based on mole fraction
$\Delta H^x_{w \to humic}$	Enthalpy changes based on mole fraction
$\Delta S^x_{w \to humic}$	Entropy changes based on mole fraction
%(V/V)	Fraction by volume (in %)
ΔG	The free Gibbs energy
ΔG_{sorp}	The free Gibbs energy of the sorption process
ΔH	Enthalpy change
ΔH_{sorp}	Enthalpy changes for the sorption process
ΔS	Entropy changes
ΔS_{sorp}	Entropy changes for the sorption process
q^2_{cv}	Predictive squared correlation coefficients using leave-one-out cross –validation
1/n	Freundlich exponent
A	Solute hydrogen-bond acidity (LSER descriptor)
B_0	Solute hydrogen-bond basicity (LSER descriptor)
bias	Systematic error
c	Constant term (solvent specific free energy contribution) in LSER
C_s	Solid phase (or sorbed) concentration of sorbate
C_w	Aqueous phase concentration of sorbate/solute
DOM	Dissolved organic matter
E	Excess molar refraction (LSER descriptor)
EDA	Electron –donor -acceptor
FA	Fulvic acid
FAs	Fulvic acids
f_{oc}	Fraction of organic carbon
FTIR	Fourier transformation infrared spectroscopy
GC-MSD	Gas chromatography with mass spectrometry detector
HA	Humic acid
HAs	Humicacids
HM	Humic matter
HOCs	Hydrophobic organic compounds
HPLC	High performance liquid chromatography
HSs	Humic substances
HS-SPME	Headspace solid phase microextraction
IHOC	Ionizable hydrophobic organic compounds

List of Symbols and Abbreviations

K_d	Soil-water sorption (distribution) coefficient
K_{oc}	Partition coefficient between organic carbon and water
K_{ow}	Partition coefficient between 1-octanol and water
$K_{sorbent/water}$	Sorbent-water distribution coefficient
ln	The natural logarithm function with the base e (2.7182818)
log	Logarithm function with the base 10
LSER	Linear salvation energy relationship
MCIs	Molecular connectivity indices
me	Mean error
mne	Maximum negative error
mpe	Maximum positive error
N	Number of data
n.a.	Not available
n.d.	Not detected
nd-SPME	Negligible depletion solid phase microextraction
OECD	Organization for Economic Co-operation and Development
OM	Organic matter
PAHs	Polycyclic aromatic hydrocarbons
PCBs	Polychlorinated biphenyls
QSAR	Quantitative structure–activity relationship
R	The "gas constant" (8.31451 J mol^{-1} K^{-1})
r^2	Linear regression correlation coefficient
rms	Root mean square error
S	Solute polarizability/dipolarityparmeter (LSER descriptor)
SD	Standard deviation
SOM	Soil organic matter
SPME	Solid phase microextraction
S_W	Aqueous solubility
T	Absolute temperature (in K)
TC	Total carbon content
TIC	Total inorganic carbon content
TOC	Total organic carbon
V	McGowan's molar volume (LSER descriptor)

List of Tables

Table	Title	Page
Tab. 1.1	Relation between soil component, receptors sites and interaction mechanisms of sorption asdescribed by MacKay et al. (2012).	10
Tab. 3.1	An example of depletion calculation for a fluoranthene probe in 20 mL vial.	48
Tab. 3.2	Elemental analysis of Aldrich HA.	51
Tab. 3.3	An example of atomic volume V for some different elements in cm^3 mol^{-1}.	63
Tab. 4.1.1	The experimental values of sorption coefficients for different organic substances at 25 ± 2 °Cand pH 7 ± 0.2 (f_{oc} of used Aldrich HA is 0.395 g/g).	69
Tab. 4.1.2	Results of linear regression of mean exp. log K_{oc} in literature versus exp. log K_{oc} in the current study.	72
Tab. 4.1.3	Results of linear regression of exp. log K_{oc} versus exp. log K_{oc} in literature for different chemical domains.	73
Tab. 4.1.4	Results of linear regression of exp. log K_{oc} versus calc. log K_{oc} using different methods.	77
Tab. 4.1.5	Results of linear regression of log K_{oc} versus log K_{ow} and comparison with similar models in literature.	89
Tab. 4.1.6	The multilinear regression statistics of phase parameters using LSER (experimental descriptors by Abraham et al. 1987, 1994a, 1994b, 1999 and 2004).	94
Tab. 4.1.7	The multilinear regression statistics of phase parameters using LSER (calculated descriptors by Platts et al. 1999).	95
Tab. 4.1.8	Results of linear regression of log K_{oc} modeling for different chemical classes.	96
Tab. 4.1.9	Comparison with the sorbent specific descriptors in literature.	97
Tab. 4.2.1	The experimental values of sorption coefficients for selected subset at 25±2 °C and at different pH values (f_{oc} of used Aldrich HA is 0.395 g/g).	103
Tab. 4.2.2	Results of linear regression of calc. log K_{oc} versus exp. log K_{oc} using different methods in literature.	110
Tab. 4.2.3	Results of linear regression of log K_{oc} versus log K_{ow} at different pH for non-ionogenic compounds, acids and bases.	114

Table	Title	Page
Tab. 4.2.4	The multilinear regression statistics of phase parameters using LSER (experimental descriptors by Abraham et al. 1987, 1994a, 1994b, 1999 and 2004).	123
Tab. 4.2.5	The multilinear regression statistics of phase parameters using LSER(calculated descriptors by Platts et al. 1999)	124
Tab. 4.2.6	Results of linear regression of suggested pH-log K_{oc} dependent prediction methods for weak acidic and basic compounds.	128
Tab. 4.2.7	Statistical performance of comparison between the experimental log K_{oc} data in the present study andcalculated data by Franco and Trapp model (2008).	129
Tab. 4.2.8	Statistical parameters of log K_{oc} prediction methods for all compounds and unionized fractions.	132
Tab. 4.3.1	The experimental values of sorption coefficients for different organic substances at pH7±0.20 and at different temperatures (f_{oc} of used Aldrich HA is 0.395 g/g).	137
Tab. 4.3.2	The experimental values of thermodynamic parameters of probe compounds at pH7± 0.20 and at different temperatures.	148
Tab. 4.3.3	Results of linear regression of calc. log K_{oc} versus exp. log K_{oc} using different methods in literature.	152
Tab. 4.3.4	Results of linear regression of log K_{oc} versus log K_{ow} at different temperatures.	154
Tab. 4.3.5	The multilinear regression statistics of phase parameters using LSER (experimental descriptors by Abraham et al. 1987, 1994a, 1994b, 1999 and 2004).	157
Tab. 4.3.6	The multilinear regression statistics of phase parameters using LSER (calculated descriptors by Platts et al. 1999).	157
Tab. 4.3.7	Statistical parameters of log K_{oc} prediction methods at different temperatures.	157

List of Figures

Figure	Title	Page
Fig.2.1	Freundlich isotherms.	21
Fig. 2.2	Sorption isotherms.	22
Fig. 2.3	Simple multiphase environmental systems.	32
Fig. 2.4	Presence of neutral or ionic form in solution according to the pH and pK_a. [AH] and [BH$^+$] are the protonated forms, [A$^-$] and [B] the dissociated forms of the acidic and basic compounds respectively (Kah et al. 2007).	34
Fig.3.1	The model structure of HA (Stevenson 1982).	39
Fig. 3.2	SPME different fibres. Red-100 µm PDMS, white- 85 µm PA, blue- 65 µm PDMS-DVB.	40
Fig. 3.3	Calibration plot of fluoranthene in solvent (2-propanol) at 25 ±2 °C.	47
Fig. 3.4	Calibration plot of fluoranthene in aqueous phase (water) in phosphate buffer pH7±0.2 at 25 ±2 °C.	47
Fig. 3.5	(a) Direct and (b) Headspace modes of SPME (Pawliszyn 2001, Krutz et al. 2003 and ter Laak et al. 2005).	48
Fig. 3.6	Equilibrium time plot of fluoranthene and atrazine in different pH values.	49
Fig. 3.7	Sorption isotherm of fluoranthene, from the Freundlich equation (Xing et al. 2008).Log C_s = Log K_f + n. LogC_w, n is Freundlich exponent >=1.046 it means sorption will be enhanced at higher concentrations of sorbate, K_F is Freundlich constant = $3.06 \cdot 10^{-5}$.	50
Fig. 3.8	Chromatogram of terbuthylazine at t_R 11.33min after 15 min SPME (85µmPolyacrylat) via headspace mode in room temperature (25 ± 2°C) for aqueous solution (181.1 µgL^{-1} in buffer solution of pH = 7± 0.2) with *SCAN* mode GC-MSD.	55
Fig. 3.9	Chromatogram of electron ionization (EI)-mass spectra of terbuthylazine at t_R 11.33 min with *SCAN* mode GC-MSD, m/z was (214, 173, and 229) for the molecular target ion and qualifier ion peaks (NIST98).	55
Fig. 3.10	Chromatogram of simetryne at t_R 12.14 min in a reference as well as a sorption experiment with 600 mgL^{-1}humic acid after 15 min SPME (85µm PA) via HS-SPME in room temperature (25 ± 2°C) for aqueous solution (144.9 µgL^{-1} in buffer solution of pH7± 0.2) with *SIM* mode GC-MSD.	56

List of Figures and Schemes

Figure	Title	Page
Fig. 3.11	Graphical plot of data for simetryne test series in 25 ± 2°C. Black lineis linear line. Green lineis limit of detection (13.5 µgL^{-1}) andblue lineis limit of quantitation (44.9 µgL^{-1}) both calculated from the calibration curve according to DIN 32645, (2011-8). Red lineis confidence interval 95% (Burke 2007).	56
Fig.4.1.1	Comparison between mean values of exp. log K_{oc} in literature and exp. log K_{oc} in current study.	72
Fig. 4.1.2	Exp. log K_{oc} in literature for different chemical domains versus exp. log K_{oc} in the current study.	74
Fig. 4.1.3	Root mean squared error (rms) versus prediction methods in literature for different chemical domains. (a) PAHs, (b) Phenols, (c) R-benzene(R= H, CH$_3$, 3(CH$_3$), and CHO), (d) Organochlorines and Biphenyls, (e) Heterocyclic compounds, (f) Halobenzenes, (g) Anilines.	78
Fig. 4.1.4	Calc. log K_{oc} in literature for different chemical domains versus exp. log K_{oc} measured.(a) Fragment constant method (Tao et al. 1999), (b) KOCWIN software (log K_{ow}) – molecular topology, (c) 2D Molecular structure (Schüürmann et al. 2006), (d) Molecular connectivity indices (Sabljić et al. 1995), (e) LSER exp. input (Poole et al. 1999), (f) LSER calc. input (Nguyen et al. 2005), (g) LSER calc. input (Endo et al. 2009a – low Peat conc.), (h) LSER exp. input (Endo et al. 2009a – high Peat conc.), (i) LSER exp. input (Neale et al. 2012- Aldrich HA in literature), (j) Franco and Trapp model for unionized compounds at pH7 (Franco et al. 2008).	84
Fig. 4.1.5	Exp. log K_{oc} versus log K_{ow} in neutral water for different chemical classes.	88
Fig. 4.1.6	Results of linear regression of calc. log K_{oc} versus exp. log K_{oc} using different models at 25 ± 2 °C and pH 7 ± 0.20 for non-ionogenic, acidic and basic compounds.	98
Fig. 4.1.7	Results of linear regression of calc. log K_{oc} versus exp. log K_{oc} using different models at 25 ± 2 °C for all compounds and unionized fractions.	99
Fig.4.2.1	Comparison between mean values of exp. log K_{oc} in literature and exp. log K_{oc} in current study	102
Fig. 4.2.2	Exp. log K_{oc} versus log K_{ow} for non-ionogenic compound, acids and bases at different pH values.	111

Figure	Title	Page
Fig.4.2.3	Log K_{oc} versus exp. log K_{ow} for non-ionogenic compounds, acids and bases at pH 4, 7 and 10.	112
Fig. 4.2.4	Log K_{oc}- log K_{ow} correlation for acids and bases for all compounds and unionized fractions. (a) acids at pH4, (b) bases at pH4, (c) acids at pH10 and (d) bases at pH10.	115
Fig. 4.2.5	Exp. log K_{oc} versus pH of HA for non-ionogenic compounds.	117
Fig.4.2.6	Exp. log K_{oc} versus pH of HA for acids.	118
Fig. 4.2.7	Exp. log K_{oc} versus pH of HA for bases.	120
Fig. 4.2.8	Results of linear regression of calc. log K_{oc} versus exp. log K_{oc} using different models at 25±2 °C for non-ionogenic compounds, acids and bases.	131
Fig. 4.3.1	Van't Hoff plot at different temperatures. r^2 is correlation coefficient, a is slope, b is intercept, straight line is significant correlation and dashed line is insignificant correlation.	144
Fig.4.3.2	Exp. log K_{oc} versus log K_{ow} in different temperatures.	155
Fig. 4.3.3	Results of linear regression of calc. log K_{oc} versus exp. log K_{oc} using different prediction methods at different temperatures.	158

List of Schemes

Scheme	Title	Page
Scheme 3.1	The chemical structure of probe compounds.	44
Scheme 3.2	SPME fibre assembly (Pawliszyn 2001).	40
Scheme 4.1	Summary of the results of temperature-log K_{oc} data dependence based on mole fraction calculations.	141

1. Introduction

1.1 Scope of the problem

Sorption as key process for the behaviour of xenobiotics in soil, the understanding of sorption process is very essential to discover the behaviour of organic soil contamination. It plays important roles in many physical, chemical and biological processes in different soils. For instance, the mobility of chemicals and their transport in soil, sediment and their accumulation in terrestrial and aquatic ecosystems, is all affected by sorption capacities of organic compounds to solid components. If the soils have high sorption capacities the transport of organic chemicals by infiltrating water is extremely low. However if the soils have low sorption capacity, then part of organic chemicals will leach down through soil profiles and go to ground water. The leachability of soil depends on the hydrophobicity of the organic contaminant, organic carbon content and the pore water. Once the organic compounds have reached the aquifer they will be affected by their sorption to aquifer materials and might be less spreading in surface water.

It is known that the sorption process has effects in other processes, such as, it limits the biodegradation and limits the toxic effects of pollutants, as controlling the freely dissolved fraction of pollutants that can accumulate in the microorganisms. Therefore, the dissolved and sorbed amounts of organic pollutants may be exposed to different environmental conditions (e.g. pH, light and temperature). Moreover the determination of sorption quantity is prerequisite for risk assessments. It is obvious that the investigation of sorption mechanisms of organic chemicals interactions will improve the predictions of environmental fate of their chemicals in the future.

The bioavailability of a soil chemical is considered the result of a series of dynamic processes, such as advection, sorption/desorption, degradation, volatilization, and uptake by microorganisms or plants roots or soil fauna. If chemicals are in contact with soil or sediments, however, sorption reduces bioavailability. The extent of sorption and sequestration is influenced by the properties of both chemical and soil. The soil sorption can be experimentally determined in the form of the partition coefficient of organic chemicals in water-soil systems K_d. It provides information about the bioavailable concentration of the contaminants (freely available concentration for microorganism). The total concentration of DOM in natural water is also one of the main factors controlling the bioavailability of highly hydrophobic xenobiotics ($K_{ow} > 10^4$). DOM reduces the bioavailability of pollutants, the magnitude of the decrease being related to the binding between the contaminant and the organic matter. Data regarding sorption, bioavailability and bioaccessibility processes is essential for quantitative and qualitative ecological risk assessment (Peijnenburg 2004, and Katayama et al. 2010).

1.2 Sorption of xenobiotics to soils and sediments

The soil sorption process has been affected normally by SOM content, but for polar and ionic molecules, the sorption can occur mainly through clay minerals. The surface of soil can play an important role as a sorbent, particularly if the organic matter fraction is low, however, the contribution of mineral surface in sorption is influenced by the total carbon content. The sorption to mineral surface can occur due to electrostatic interaction, hydrophobic interactions, and specific bonding reactions at the surface such as cation and anion exchanges. Ligand exchange reactions are the most important process in the sorption of DOM on mineral phases. Sorption of DOM to mineral surface through the formation of strong chemisorptive bonds via surface complexation of functional groups to metal oxides and hydroxides in soils can be an important mechanism in the preservation of SOM (Mader et al. 1997, Kaiser et al. 2000, and Oren et al. 2012). The magnitude of sorption coefficient normalized to organic carbon (log K_{oc}) for a specific pollutant in a water-soil system has been shown to be related to partition coefficient between 1-octanol and water (log K_{ow}) of the pollutant, the organic carbon content of the solid phase onto which sorption occurs as variation of characteristics of the humic material. Log K_{oc} values decreased with increasing polar group content of the HA. Literature has shown the influence of dissolved humic and fulvic acids on water solubility enhancement and resulting effects on the partition coefficients of organic chemicals in soils. It was observed that K_{oc} values of solutes with soil-derived HAs are approximately 4 times greater than with soil FAs and 5 - 7 times greater than with aquatic HAs and FAs (Krop et al. 2001). For non-polar and low polar substances the organic matter content of the soil is the predominant soil constituent influencing sorption (Krop et al. 2001 and Kördel et al. 1997).

Highly hydrophobic organic compounds such as PAHs which are practically insoluble in water tend to absorb onto other non-aqueous phases either through hydrophobic interaction, when the non-aqueous phase is a nonpolar compound, or through conjugate bonding, when the non-aqueous phase is a polar compound. The hydrophobic sites of HSs represent an important phase for PAHs. The interactions between PAHs and dissolved HSs may not only enhance the concentration of hydrophobic organic compounds in the aqueous phase, but may also effectively desorb pollutants. If the DOM is associated with the aqueous phase, PAHs will have a tendency to bind to the DOM and stay in the solution, potentially facilitating their subsurface transport, but also if it is associated with a soil matrix, PAHs may sorb onto the soil phase and its transport will be retarded. The range of K_{oc} values for apolar substances such as PAHs determined with different soils is usually within a factor of 10 (Krop et al. 2001, Kördel et al. 1997, and Kipka et al. 2011).

1.3 Sorption of humic acid

The sorption of xenobiotics to HA as model for natural sorbent depends on the properties of the sorbent and the physicochemical characterization of organic chemicals in addition to the environmental conditions of the media (Krop et al. 2001).

The commercial HA possess a higher log K_{oc} than neutral ones (Krop et al. 2001). The slope of the log K_{oc}–log K_{ow} correlation has shown that the HSs obtained from water samples possess a lower sorption capacity than those from soils and sediments as shown in the following equations (1.1-1.4). Aldrich HA may have a lower functional variability than natural DOM (Krop et al. 2001).

$$\log K_{oc(commercial\ HA)} = (0.68 \pm 0.14) \log K_{ow} + (1.4 \pm 0.7) \qquad (1.1)$$
$$r^2 = 0.65,\ n = 44,\ F = 80$$
$$\log K_{oc(sediment\ HA)} = (0.66 \pm 0.44) \log K_{ow} + (1.3 \pm 2.5) \qquad (1.2)$$
$$r^2 = 0.57,\ n = 11,\ F = 11$$
$$\log K_{oc(soil\ HA)} = (0.53 \pm 0.25) \log K_{ow} + (1.9 \pm 1.3) \qquad (1.3)$$
$$r^2 = 0.34,\ n = 36,\ F = 17$$
$$\log K_{oc(water\ HA)} = (0.34 \pm 0.20) \log K_{ow} + (2.9 \pm 1.3) \qquad (1.4)$$
$$r^2 = 0.33,\ n = 24,\ F = 11$$

Literature data shows that the sorption of aquatic HAs is two or three times higher than that of aquatic FAs and both are smaller than the sorption of HAs and FAs from soils. Log K_{oc} data of FAs are lower than those of HAs from the same sample due to lower molecular weight and lower hydrophobicity (Krop et al. 2001). It was expected to find a less hydrophobic character for neutral HA obtained from water samples. Hydrophobic contaminants with a log K_{ow} in the range 3 - 8 show a total variation of 4 log units in log K_{oc} (Krop et al. 2001). Both non-polar and polar interactions are thought to be responsible for the self-aggregation of humic material. Strongly hydrophobic contaminants such as PAHs would stay mainly in the DOC phase for thermodynamically reasons. Polar compounds tend to remain more in the DOC phase than in the soil organic phase due to low possibilities of polar interactions in the soil organic phase (Krop et al. 2001).

1.4 Selection of probe compounds

The term **xenobiotics** are used for different classes of pollutants and their effect on the biota. They enter the environment through human activity and may occur in amounts high enough to be harmful to living organisms. The environmental chemicals can be dangerous and may be responsible for acute or chronic health problems after accumulation or molecular change. The present study focuses on one of the fate processes (partitioning) of some xenobiotics in water-soil environment and aims to answer the question how the quantification of their fate in the environment may be used to predict it in the future. In the current study, different categories of organic chemicals were selected and represented for different types of xenobiotics such as polycyclic aromatic hydrocarbons and their derivatives (**PAHs**) as partly carcinogenic and teratogenic compounds. **Phenols** may possess estrogenic or endocrine disrupting activities. **Organochlorines** have significant reactivity and toxicity to plants, animals and humans. **Triazines** are a suspects of teratogen and estrogen disrupting activities

1.5 Literature review of sorption process

Many studies are available on the sorption of organic chemicals in soils (among the most recent: MacKay et al. 2012, Neale et al. 2012, Bronner et al. 2010a, b, and Endo et al. 2008, and 2009 a, b, c, Franco et al. 2008 and 2009 and Schüürmann et al. 2006, and 2007). These mainly cover the behaviour of non-polar organic compounds in soils, which is now relatively well understood, while the sorption of ionizable compounds still needs more investigation. Although the information of sorption process are available, there is still much debate regarding the underlying mechanisms and the approaches to describe and predict variation in sorption with variation of chemical domain, temperature, and pH. Numerous articles reported results about the sorption to HAs and soils and sediments in the past fifteen years (Delle Site 2001, Krop et al. 2001, Doucette 2003, Xing et al. 2008 and MacKay et al. 2012). In the next paragraphs, some recent studies were selected to present the state of knowledge on the particular behaviour of sorption of polar and non-polar organic compounds.

1.5.1 Sorption of xenobiotics in the aquatic environment

It is known that during the phase transfer of the compounds from aqueous solution to the solid phase, there are different types of interaction forces affecting the association of xenobiotics to the solid matter. There may be van-der-Waals interaction forces (e.g., London forces, Keesom forces and Debye forces) and, if speciation or protonation takes place, there can be new Coulomb forces, and interactions between opposite charges will occur (von Oepen et al. 1991b, Schüürmann et al. 2007 and Vitha et al. 2006). Two or more mechanisms may take place simultaneously on the surface of HSs as organic matter sorbent. Such mechanisms may include van-der-Waals attraction, hydrophobic bonding (hydrophobic partitioning), hydrogen bonding, charge transfer, ion exchange, ligand exchange, cation bridging and/or bound residues (Kah 2007 and von Oepen et al. 1991a).

There has been evidence that carbonaceous geosorbents (black carbon, humin, kerogen and coal) control the soil/sediment sorption of nonpolar hydrophobic organic compounds such as polychlorinated biphenyls, chlorinated benzenes and polycyclic aromatic hydrocarbons (Zhu et al. 2005a and Cornelissen et al. 2005). Presence of dissolved HA might significantly impact on xenobiotics adsorption via blockage of microspores of carbonaceous adsorbents and competitive adsorption for surface sites (Ji et al. 2011).

In the published literature discussing the dual-mode concept of organic matter such as solid state of OM, it is depicted as an amalgam of rubbery and glassy phases. Both phases have dissolution domains, but the glassy phase also has holes where specific interactions occur. Holes are conceptualized as cavities or microvoids and the binding is possibly analogous to host-guest inclusion complexes. In addition to some findings of nonlinearity, it has been shown that the nonlinearity and competitive effects increase with the degree of condensation of HSs as well as with equilibrium time due to the activation of hole-filling and/or the fact that holes are not sufficiently capable of accessing the less accessible regions in HSs (Xing et al. 1997).

A new view of HSs structure has been described by Sutton et al. (2005), whereas HSs are collections of diverse, relatively low molecular mass components forming dynamic associations stabilized by hydrophobic interactions and hydrogen bonds. These associations can be organized into micellar structures in aquatic environment. Humic components have a contrasting molecular motional behavior also comprise any molecules intimately associated with HSs.

Liu et al. (2008) and Qu et al. (2008) have demonstrated that the solute molecular structure and soil nature in addition to solute hydrophobicity and organic carbon content of soil control the retention of organic contaminants by soils. They have found both original and normalized sorption data of PAHs to be greater than those of chlorobenzenes with close hydrophobicities to the Shenyang, Suzhou and Kaifeng soils, due to π-π electron-donor-acceptor (EDA) interactions of PAHs with π-acceptor structure. They observed enhancement of sorption and aqueous solubility of PAHs attributed to cation–π interactions between ammonium cations and PAHs, while chlorobenzenes are incapable of such interactions (Qu et al. 2008).

The sorption capacity of various plant cuticular fractions, including the dewaxed-hydrolyzed residue, for nonpolar and polar organic pollutants has been investigated by Chen et al. (2005). For the aliphatic-rich sorbents, no correlation of K_{oc} with either aromatic or aliphatic components was found because the polarity and accessibility played a regulating role in the sorption of organic contaminants. Moreover, there was a significant effect of polarity and accessibility of organic matter in the uptake of nonpolar and polar organic contaminants by regulating the compatibility of sorbate to sorbent. The K_{oc} values decreased with increasing sorbent polarities, and the decreasing rates of sorption were dependent on organic pollutants' properties and their corresponding sorption mechanisms (partition or specific interaction).

The compatibility between HOCs and SOM needs to be considered in HOCs predictive and risk assessment models. Chefetz et al. (2009) have investigated the influence of aliphatic and aromatic domains on sorption for HOCs. It was observed that (i) aliphatic structures must be considered in the evaluation of HOC-sorption processes in the environment; (ii) neither aromaticity nor aliphaticity of SOM alone can be used to predict the sorption affinity of sorbents having wide and diverse properties.

K_{oc} plays an important role in risk assessment; therefore the appropriate predictive tools of K_{oc} are very essential. The classical approach and most commonly used method is to predict K_{oc} from log K_{oc} - log K_{ow} correlations (K_{ow} is octanol/water partition coefficient), however this model is applicable only for non-polar hydrophobic organic compounds (Karickhoff 1981, Sabijić et al. 1995, Gawlik et al. 1997, and Doucette 2003). The sorption of most ionizable hydrophobic organic compounds is affected by soil solution's pH and the pK_a. When the neutral fractions of acids and bases are lowered by dissociation and protonation, then the sorption behavior will be changed and lowered by decreasing pK_a of the acid form of the solute (Gawlik et al. 1997, Doucette 2003 and Kah et al. 2006). In this case K_{ow} becomes an insufficiently reliable predictor of K_{oc}.

The prediction of log K_{oc} from molecular structure is very important approach of QSAR models. An existing 2D molecular structure model (Schüürmann et al. 2006) is a multilinear model containing the variables molecular weight, bond connectivity, molecular E-state, an indicator for non-polar and weakly polar compounds, and 24 fragment corrections representing polar groups. A molecular connectivity indices model (Sabljic´ et al. 1995) is employing 19 different regression equations based on log K_{ow} and one $^1\chi$ based regression to calculate log K_{ow} from molecular structure using KOWWIN. A fragment constant model (Tao et al. 1999) is consisting of 74 fragment values and 24 structural correction factors.

Recently Zhu et al. (2005b) have found that as an application of the LSER approach, (i) sorption of a calibration set of nonpolar compounds (aromatic and aliphatic hydrocarbons and chlorinated hydrocarbons) to the natural sorbents is well described by a combination of hydrophobic and dipolarity/polarizablility effects. For the apolar set, dipolarity/polarizablility contributes ~ 15 - 40 % (2 - 8 % for cyclohexane) of sorption free energy. Dipolarity/polarizablility effects increase with the degree of chlorination for aliphatic compounds. For aromatic compounds, dipolarity/polarizablility effects increase with fused ring size but do not vary with degree of chlorination and chlorine substitution pattern. (ii) Solutes molecules fill sites of progressively greater hydrophilic character. (iii) The energy penalty for cavity formation in the solid decreases with concentration due to plasticization and greater intermolecular contact. (iv) Sorbent aromatic content mostly controls dipolarity/polarizablility interactions.

Poole et al. (1999) and Nguyen et al. (2005) have proposed and calibrated the LSER equation for sorption to soil organic matter based on experimental literature data. Niederer et al. (2007) have provided both experimental data and calibrated equation of LSER of sorption data for 90 diverse organic chemicals from air to 10 hydrated HAs and FAs. The proposed models can account for van-der-Waals as well as H-donor/acceptor interactions between sorbate and sorbent.

In more recent LSER studies, published by Endo et al. (2008 and 2009 a, b, c), some findings showed that (i) Pahokee peat soil sorption gradually shifts from adsorption on more aromatic and/or polar sites (carbonaceous surfaces or glassy domain of organic matter) to absorption in sites of less aromaticity/polarity (organic matter or rubbery part of organic matter). (ii) Two types of nonlinear sorption by organic matter using LSER-based approaches have been proposed: one controlled by compound-independent factors and the other by the dipolarity/polarizability property S of the compound. (iii) An empirical relationship between the nonlinearity (expressed by the Freundlich exponent, $1/n$) in peat and the sorbate dipolarity/polarizability property S was reported (Endo et al. 2008).

$$1/n = (0.97 \pm 0.01) - (0.27 \pm 0.02)\, S \qquad (1.5)$$

$$r^2 = 0.90,\ \text{for Pahokee peat}$$

Therefore, it would be possible to estimate log K_{oc} values at varying concentrations based on a log K_{oc} value determined at a single concentration.

Goss et al. (2005) have replaced E-descriptor by L-descriptor (the logarithm of hexadecane/air partition constant at 25 °C in LSER equation) because of some disadvantages of E-descriptors for solid compounds, where it cannot be calculated from experimental refractive indices (Tülp et al. 2008). Therefore an empirical method for estimation is required.

Bronner et al. (2010a) have established a new LSER model for soil-water partition coefficient for pesticides and other multi-functional organic chemical. The model was calibrated with data for 79 polar and nonpolar compounds that cover a very wide range of the relevant intermolecular interactions.

$$\log K_{oc} = (0.81 \pm 0.08)\, E - (0.61 \pm 0.11)\, S - (0.21 \pm 0.14)\, A - (3.44 \pm 0.18)\, B + (2.99 \pm 0.11)\, V - (0.29 \pm 0.12) \qquad (1.6)$$

$$r^2 = 0.921,\ n = 79,\ SE = 0.25$$

Neale et al. (2012) have measured $K_{DOC\text{-}w}$ for a range of non-polar and polar compounds with Suwannee River FA using headspace and SPME methods. The selected chemicals represented a range of properties including van-der-Waals forces, cavity formation and hydrogen bonding interactions. LSER approach was applied, and a difference between experimental and calculated values of less than 0.3 log units was found, indicating that the calibrated LSER gives a good indication of micropollutant interaction with FA. The LSER approach was applied for Aldrich HA using $K_{DOC\text{-}w}$ values collected from literature. Both experimental and LSER calculated $K_{DOC\text{-}w}$ values for Aldrich HA were found to be greater than Suwannee River FA by one order of magnitude. This difference was attributed to the higher cavity formation energy in Suwannee River FA. Experimental and LSER calculated $K_{DOC\text{-}w}$ values were compared for halogenated alkanes and alkenes, including trihalomethane disinfection by-products. It was observed that sorption to DOC is not an important fate process for these chemicals in the environment.

For Suwannee River FA

$$\log K_{DOC\text{-}w} = (0.63 \pm 0.19) E - (0.63 \pm 0.28) S + (0.05 \pm 0.26) A - (2.48 \pm 0.26) B + (2.86 \pm 0.26) V$$
$$-1.21 (\pm 0.36) \quad (1.7)$$
$$r^2 = 0.845, n = 34, SD = 0.24$$

For Aldrich HA in literature

$$\log K_{DOC\text{-}w} = (0.59 \pm 0.08) E - (0.52 \pm 0.12) S + (0.63 \pm 0.19) A - (3.40 \pm 0.16) B + (3.94 \pm 0.16) V$$
$$-(0.85 \pm 0.16) \quad (1.8)$$
$$r^2 = 0.971, n = 52, SD = 0.29$$

MacKay et al. (2012) have shown that polyfunctional ionogenic compounds are sorbed to environmental solids at multiple receptor sites via multiple interaction mechanisms. The presented unique mapping of sorbate structural moieties, sorbent receptor sites, and sorption mechanisms is used to advance mechanism-specific probe compounds for cation exchange and surface complexation/cation bridging for quantifying the relevant site abundance and baseline sorption free energy (see Table 1.1).

Table 1.1 Relation between soil component, receptors sites and interaction mechanisms of sorption as described by MacKay et al. (2012).

Soil Component	Receptor Site	Interactions Mechanisms
Organic Matter	negative charge	cation exchange, cation bridging
	non-polar domain	hydrophobic partitioning
	polar domain	electron donor acceptor (EDA)-specific type hydrogen bonding
Aluminosilicates	negative charge	cation exchange, cation bridging
	surface bound Fe and Al	surface complexation via ligand exchange
	positive charge	anion exchange
	polar domain	electron donor acceptor (EDA)-specific type hydrogen bonding
Metal oxides	negative charge	cation exchange, cation bridging
	surface bound Fe and Al	surface complexation via ligand exchange
	positive charge	anion exchange
	polar domain	electron donor acceptor (EDA)-specific type hydrogen bonding

1.5.2 Influence of pH on sorption process

The ionizable hydrophobic organic compounds possess either basic or acidic functional groups. They can be partially ionized (dissociated) within the range of neutral soil pH. The adsorption of ionizable compounds in the soil is strongly influenced by pH. This pH dependence derives mainly from the different proportions of ionic and neutral forms of compounds present at pH 4 - 10 depending on the degree of dissociation (Kah et al 2007, Franco et al. 2008, and 2009) (i.e., the ratio of neutral species to the total species). The sorption of most ionizable hydrophobic organic compounds is affected by soil solution's pH and the pK_a. When the neutral fractions of acids and bases are reduced by dissociation and protonation, then the sorption behavior will be changed and lowered by decreasing pK_a of the acid form of the solute (Doucette 2003, Kah et al 2007, Franco et al. 2008, and 2009). As such, this set of data is important for an understanding of partitioning, bioconcentration and associated ecotoxic effects of ionogenic compounds.

In the classical molecular hydrophobic prediction model for sorption of organic compounds into organic matter, Karickhoff (1981) showed a linear relationship between log K_{ow} and log K_{oc}. However, the model is limited and only suitable for neutral compounds, and it failed for ionisable compounds. It is not suitable for the prediction of the sorption behavior of ionogenic compounds. Few studies (Franco et al. 2008, and 2009 and Tülp et al. 2009) have investigated the sorption of ionizable compounds and the log K_{oc} - pH dependence.

Kah et al. (2008) have studied the adsorption behaviors of ionizable pesticides in soil at different pH, the impact on the charge on the soil surface, and also the dissociation and prorogation of pesticides depending on their pK_a values. It has been found that the adsorption of ionizable pesticides is stronger in lower pH soils and high organic carbon content, and that the influence of these two parameters is higher for acids than for bases. Moreover, the neutral and ionic forms of ionizable compounds have different polarities, because their ratio varies with pH, and therefore the hydrophobicity of ionizable pesticides is pH-dependent. The hydrophobicity decreases with increasing pH due to the increase of anionic species, meanwhile the hydrophobicity of bases increases with pH since the neutral form is dominant at pH > pK_a.

Recently, the literature has focused on advanced studies of the nature of binding with organic matter for both polar and non-polar organic compounds. Tülp et al. (2009) have studied the sorption of the neutral and anionic species of 32 diverse organic acids belonging to nine different chemical groups to SOM. K_{oc} data have been determined using HPLC retention volumes on a column packed with peat, at three Ca^{2+} concentrations and over a pH range of 4.5 - 7.5. K_{oc}, of both neutral and anionic species increased with increasing molecular size and decreased with increasing polarity. It was found that the non-ionic interactions govern the partitioning of both the neutral and anionic species into SOM, while the electrostatic interactions of the anionic species with SOM are complex and not well investigated, in particular regarding the role of the type of acidic functional group.

Neale et al. (2009) have investigated the interaction of steroid hormone with environmentally relevant concentrations of Aldrich HA, alginic acid and tannic acid using SPME. It has been found that the effect of pH on sorption influences their fate and bioavailability. For HA and tannic acid, sorption was strongest at acidic pH when the bulk organic matter was in a non-dissociated form, and it decreased when they became partially negatively charged. Furthermore, at acidic and neutral pH the strength of partitioning has been influenced by hormone functional groups content. Otherwise, in alkaline pH conditions, when the bulk organics were dissociated, sorption decreased, although the non-dissociated hormones have shown greater sorption to HA at pH 10 compared to the partially deprotonated hormones.

Different studies (Doucette 2003, Kah et al 2007, Franco et al. 2008, and 2009, Tülp et al. 2009 and Bronner et al. 2010) have shown the dependence of organic compound sorption on their polarity and structure. Phenols are weakly polar hydrophobic ionizable organic solutes in groundwater. Their toxicity is influenced by groundwater pH. Ionic species are more soluble in aqueous phase and more mobile in the subsoil, while neutral species are more toxic and have a high affinity for binding with organic matter. Fiore et al. (2009) have found that the sorption of phenols has been influenced by some predominant factors such as f_{oc}, cation exchange capacity, ground water pH and pollutants' hydrophobicity. The highest f_{oc} and the presence of the nitro-group of phenols have shown the highest amounts of pollutant transferred to the solid phase. The interactions between phenols could be occurring for both phenolate and neutral forms.

Chefetz et al. (2010) have studied the mechanism of interaction of the pharmaceuticals naproxen and carbamazepine with structural fractions of biosolids-derived DOM using dialysis-bag experiments at different pHs. Sorption of naproxen and carbamazepine has exhibited strong pH-dependence by the hydrophobic acid fraction. The hydrophilic acid fraction exhibited the highest sorption for carbamazepine, probably due to its bipolar character. In the hydrophilic acid fraction-naproxen system, significant anionic repulsion was observed with increasing pH. The hydrophilic base fraction contains positively charged functional groups. Therefore with increasing ionization of naproxen (with increasing pH), sorption of this fraction increased. The study highlighted the differences between K_{DOC} values and calculated values for both carbamazepine and Naproxen. It concluded that DOM fractions interact with each other and do not act as separate sorption domains.

Bronner et al. (2010b) have assumed that the organic carbon/water partition coefficient (K_{oc}) of neutral organic chemicals can be treated as a constant property that remains unaffected by the type of soil organic matter and pH in the soil solution. Sorption experiments with three different soils and one peat have been carried out using a column method. It has been found that the difference in log K_{oc} at pH values of 4.5 and 7.2 is on average < 0.06 log units for 60 chemicals on Pahokee Peat. Protonation/deprotonation of carboxylic groups in HM has no significant influence on sorption.

The effect of pH changes on sorption of phenanthrene and atrazine by HAs and HM has been studied by Wang et al. (2011). It has been shown that the zeta potential of HSs decreases with increasing pH. For both phenanthrene and atrazine, the sorption increases at pH 4, but it decreases with increasing pH. Atrazine would be protonated at low pH, and this protonation favors increased sorption by HSs via ionic interactions. In fact, the organic matter in HAs and HM has a higher abundance of hydrophobic sites at lower pH values than at higher ones. As an apolar compound, phenanthrene is unable to be sorbed to HAs and HM through ionic interactions at both lower and higher pH values. The organic matter in HAs and HM may have higher nonspecific interactions with phenanthrene at lower relative to higher pH values, due to its more hydrophobic nature.

The prediction model of soil sorption of organic electrolytes based on pK_a was proposed by the Franco and Trapp theoretical model (Franco et al. 2008 and 2009), depending on the log $K_{ow(neutral\ species)}$ and pK_a of selected solutes. This new approach can be used to predict the soil sorption of ionisable substances in soil and sediments. The prediction model considers the sorption of electrolytes by treating neutral and ionic fractions separately and can provide good estimates of the log K_{oc}. The second proposed model considers speciation as a function of soil pH and species-specific partitioning equilibrium to predict pH dependent K_d values of organic acids. For bases sorption, the model fits only at pH 4.5 as shown in Appendix 9.18.

1.5.3 Influence of temperature on sorption process

Sorption can spontaneously occur if the free energy between the organic solute and DOM is negative (Doucette 2003 and ten Hulscher et al. 1996), in the case of HA then the free energy of products (organic solute-HA complex) is lower than that of the reactants and mainly stable complex are formed. Many types of molecular interaction can occur such as enthalpy dependent forces (van-der-Waals interaction, hydrogen bonding, and charge transfer), while other forces are entropy dependent as well as hydrophobic bonding or hydrophobic partitioning (ten Hulscher et al. 1996, von Oepen et al. 1991a and b).

Literature data on the temperature dependence of log K_{oc} are very rare, since earlier work in literature focused on other basic physicochemical compound properties such as log K_{ow} (Paschke et al.1998 and Congliang et al. 2007), Henry`s law constant (Kühne et al. 2005 and Böhme et al. 2008) and the peptides–membrane partition coefficient by both experimental and calculation methods (Wong et al. 1998).

Temperature plays a significant role in the sorption process of organic pollutants, and its influence on the sorption equilibrium is indicative of sorption energies and mechanisms. It is expected that the sorption at low temperatures is higher than the sorption at high temperatures, due to a decrease in water solubility and increase in the desorption partition constant (Tremblay et al. 2005, Niederer et al. 2006, Chen et al. 2007, Jia et al. 2010 and Wang et al. 2011).

Ten Hulscher et al. (1996) have shown that the equilibrium sorption of most organic solutes decreases with increasing temperature, and that there is no effect of temperature on sorption equilibrium for other solutes of short equilibrium time attributed to slow equilibrium to solid phase increases and fast equilibrium with water decreases can occur simultaneously. When these two contributions are equal in magnitude but with opposite sign, they cancel each other out and result in very little or no net temperature dependence. Tremblay et al. (2005) have observed that sorption at different temperatures is a slightly exothermic process with calculated enthalpies between -4.5 and -13.2 kJ/mol. The impact of temperature and variations can be mainly attributed to changes in the contaminant's water solubility.

Niederer et al. (2006) have applied the thermodynamic cycle in the HA/water/air system to calculate the sorbent descriptors of HA–organic carbon/water from log $K_{Humic-air}$ at different temperatures using LSER approaches.

Chen et al. (2007) have investigated temperature sorption-dependence of phenanthrene and anthracene from 10 °C to 40 °C. They it showed that sorption of phenanthrene and anthracene onto sediments decreased when temperature increased. The magnitude of the decrease in sorption was attributed to the increased desorption rate constant, solubility, and heterogeneities of sediments.

Jia et al. (2010) have studied the sorption of perfluorooctane sulfonate (PFOS) on HA. Sorption capacity doubled when the temperature was increased from 5 to 35 °C, with thermodynamic calculations indicate that the sorption was a spontaneous, endothermic, and entropy driven process. Wang L et al. 2011 have described the sorption of five PAHs on three heterogeneous sorbents at a temperature range of 5 - 35 °C. Strong sorption was observed at lower temperatures. K_d increased 2 - 5 times as the temperature decreased due to changes in the water solubility of PAHs. Enthalpy changes (ΔH) for the sorption process were calculated, and the values were observed to be negative for all the interactions, suggesting that the exothermal sorption of PAHs inversely dependents on temperature.

1.6 The importance of the topic for the Egyptian environment in the future

Egypt has developed a national implementation plan of the Stockholm convention on persistent organic pollutants (POPs) in 2002 – (modified 2009). Persistent organic pollutants are chemicals that remain intact in the environment for long periods, become widely distributed geographically, accumulate in the fatty tissue of living organisms and are toxic to humans and wildlife. POPs circulate globally and can cause damage wherever they travel. They have toxic properties, resist degradation, are subject to bioaccumulation and are transported through air, water, soil and sediment, and migratory species across international boundaries and deposited far from their place of release, where they accumulate in terrestrial and aquatic ecosystems.

Egypt signed the convention on persistent organic pollutants in May (2002) in Stockholm, Sweden, joining forces with more than 100 other countries to support and implement the United Nations treaty. However, Egyptians are still at risk from POPs that are persistent in the environment, from POPs produced in Egypt, and from POPs released elsewhere and transported to the Egyptian environment by air or by the Nile water. Therefore, this present study has a good contribution for the environmental research in Egypt. It is a part of a scientific approach to protect the Egyptian environment especially in the field of pollution reduction and ecological risk assessment.

1.7 The importance of this topic for REACH in Europe

REACH is a European Community Regulation concerning the registration, evaluation, authorization and restriction of chemical substances, which entered into force on 1 June 2007. REACH aims to improve the protection of human health and the environment through better and earlier identification of intrinsic properties of chemical substances. REACH regulation provides the management of risks of chemicals and safety information of products. At present, there is a lack of sufficient information on the hazards chemicals pose to human health and the environment. Therefore, there is a need to fill the information gap by developing improved methods and techniques for assessing the hazards and risks of these substances in order to identify and implement risk management measures to protect humans and the environment in the future.

1.8 Scope and outline of the thesis

The author conceives that the present study not only provides environmental analysis and the determination of new experimental data, but also focuses on the interpretation and the environmental applications of physicochemical compound properties, as well as mechanistic information regarding some environmental fate processes and the performance of prediction methods.

The aim of this PhD is to contribute to a better understanding and prediction of the sorption of organic chemicals from different chemical domains to HA. The main objectives of this work are:

- to review current knowledge regarding the soil sorption of polar and non-polar organic compounds by collecting and comparing with the published log K_{oc} values of the probe compounds to soils, sediments and humic acids from (1995 till 2011) and to understand which factors influence the sorption and how a different interaction mechanisms can occur between the HOC and SOM,
- to study the relevance of batch experiments companied with direct immersion and headspace modes of nd-SPME technique to determine the sorption coefficients for polar and non-polar organic compounds, validate the method and its limitation,

- to gain insights into the sorption processes of hydrophobic organic compounds under consideration of the applicability domains by determination of sorption interaction mechanisms for a set of organic compounds from different chemical classes such as (PAHs, organochlorines, biphenyls, phenols, anilines, and heterocyclic compounds) in neutral conditions at pH7 and 25 °C,
- to identify the organic matter and organic compounds properties influencing the sorption onto HA and investigate interactions between the functional groups of sorbent and sorbates in different conditions (pH and temperature variation) by determination of pH-dependence of sorption for selected organic compounds of acidic, basic and neutral character in pH 4, 7 and 10 and determination of temperature-dependence of sorption for selected organic compounds of wide range of hydrophobicity (log K_{ow} from 0.65 to 8.27) at different temperature 5, 25, and 45 °C,
- to propose approaches to predict the extent of these processes when data are not available by using QSAR approaches of log K_{ow}, LSER, mixed between them to predict log K_{oc} for each chemical domain and show their predictability performance,
- to compare the experimental data of log K_{oc} in the current study with other existing prediction methods in literature such as 2D molecular structure model by Schüürmann et al. (2006), KOCWIN software based on molecular topology, KOCWIN based on experimental database of log K_{ow}, molecular connectivity indices model by Sabljic' et al. (1995), fragment constant model by Tao et al. (1999), Franco and Trapp model (2008) depending on pK_a, and LSER models by Nguyen et al. (2005), Poole et al. (1999), Endo et al. (2009a), Bronner et al. (2010a), and Neale et al. (2012).

2. Theoretical background

2.1 Scope in sorption phenomena

The **sorption** process is defined as a surface phenomenon which may be either **ab**sorption or **ad**sorption, or a combination of the two. It is a process in which chemicals interact with and "dissolve" in or adhere to a solid phase, such as soil. Two basic processes of sorption occur via adsorption and absorption. Adsorption refers to the sorption of a chemical onto a two-dimensional surface, while absorption describes the dissolution of a chemical inside of a 3-dimensional matrix (Schwarzenbach et al. 2003).

It is of great interest to study the sorption mechanisms which are essential for assessing subsurface contaminant flux, as they significantly affect the rate at which contaminants move through soil. They explain the transfer and persistence of compartments in soil and provide predictions for the environmental fate and impact of contaminants. They play a major role in bioavailability, biodegradability, photolysis, and hydrolysis of toxicants in aquatic ecosystems. After sorption to a solid surface or matrix, the chemicals involved have been transferred into the solid or "sorbed" phase. Phase transfer into the sorbed phase can take place from aqueous, gaseous or organic phases, and has been shown to be governed by the same type of equilibrium partition coefficient (K_d) as transfer into other phases. At equilibrium, when the concentration of a contaminant in water is increased, the concentration of that contaminant in the soil/sediment will also increase by a constant factor. This linear relationship is expressed mathematically via the solute distribution coefficient, which is the ratio of the solute concentration in soil (C_s) to the solute concentration in water (C_w).

$$K_d = \frac{C_s}{C_w} \tag{2.1}$$

If the concentration of an organic contaminant in one compartment is known, either water or soil/sediment, it allows researchers to predict the concentration of a contaminant in the other compartment. As the contaminant sorption occurs predominantly by partition into the soil organic matter (SOM), it is more useful to express the distribution coefficient in terms of the organic carbon content.

$$K_{oc} = \frac{K_d}{f_{oc}} \qquad (2.2)$$

K_{oc} is the partition coefficient normalized to the organic carbon of the soil/sediment, and f_{oc} is the organic carbon fraction of the soil (or sediment). K_{oc} values can be used to assess the sorption of non-polar and polar contaminants to the OM of different soils/sediments. The soil/water partition coefficient normalized to organic carbon content in its logarithmic form (log K_{oc}) is one of the key parameters for modelling the uptake of persistent organic contaminants regarding to their mobility, bioavailability, toxicity, and fate in the soil (Doucette 2003 and Xing et al. 2008).

The sorption of contaminants into soil or sediment varies with the nature of the contaminant and the characteristics of the soil and sediment. The composition of soil and sediment includes both mineral surface and OM. The soil or sediment can be described as a dual-function sorbent, in which the mineral matter sorbs the contaminant by adsorption (the sorbed organic compounds are held on the surface of the mineral grains), while the SOM sorbs the contaminant by a partition process in which the sorbed organic compounds dissolve (partition) into the matrix of the entire SOM. In the presence of water, the mineral surface prefers to adsorb water because of their similar molecular polarities, while the SOM prefers to absorb the contaminants by partitioning. Therefore, it is important to understand the unique function of the SOM within aquatic systems and how the partition processes affect the fate of common environmental contaminants.

2.2 Sorption isotherms

The capacity for a soil or mineral to adsorb a solute from solution can be determined by an experiment called a batch test. In a batch test, a known mass of solid (S_m) is mixed and allowed to equilibrate with a known volume of solution (V) containing a known initial concentration of a solute (C_i). The solid and solution are then separated and the concentration (C) of the solute remaining is measured. The difference $C_i - C$ is the concentration of solute adsorbed. The mass of solute adsorbed per mass of dry solid, q, is given by the following equation (2.3),

Chapter 2 - Theoretical Background

$$q = \frac{(C_i - C)V}{q_s} \quad (2.3)$$

where q_s is the mass of the solid. The test is repeated at constant temperature but varying values of C_i. A relationship between the equilibrium concentration C_{eq} and q can be graphed. Such a graph is known as an isotherm and is usually non-linear. Two common equations describing isotherms are the Freundlich and Langmuir isotherms.

Linear Isotherms - The simplest form is the linear relationship between q and C_{eq} which is:

$$q = a + b\,(C_{eq}) \quad (2.4)$$

In this instance linear regression may be used to determine the slope (b) and the intercept (a), Many sorbates exhibit linear isotherms at low concentrations.

Freundlich Isotherms - A second type of isotherm exhibits increasing adsorption with increasing concentration, but a decreasing positive slope as C_{eq} increases. Many organic and inorganic compounds follow this type of sorption behavior. This isotherm is termed the "Freundlich Isotherm". It is described by the following equation (2.5),

$$q = K_f C_{eq}^{\frac{1}{n}} \quad (2.5)$$

The logarithmic form of the Freundlich equation is expressed as:

$$\log q = \log K_f + \frac{1}{n} \log C_{eq} \quad (2.6)$$

where K_f is the "Freundlich" equilibrium constant and $1/n$ is an arbitrary constant evaluated by linearizing the equation. In this case, K is called the distribution coefficient (K_d).

$1/n$ represents the relative magnitude and diversity of energies associated with sorption process,
- if $1/n = 1$: a linear isotherm is obtained from linear adsorption due to equal adsorption energies for all sites. It occurs at very low solute concentrations.
- if $n \neq 1$ and ($1/n > 1$): represents a concave, curved upward isotherm, where the sorption energy increases with increasing surface concentration may be attributed to strong adsorption of the solvent, strong intermolecular attraction within the adsorbent layers, or

penetration of the solute in the adsorbent. This sorption behaviour is common for the soil fine fractions, which have a higher total amount of associated organic matter, than for the coarse fractions.

- if $n \neq 1$ and $(1/n < 1)$: represents a convex, curved downward isotherm where the sorption energy decreases with increasing surface concentration. It occurs where the competition of solvent for sites is minimum or the absorbate is a planar molecule.

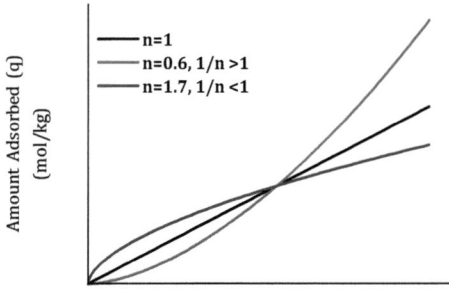

Equilibrium Concentration (C_{eq}) (mol/L)

Figure 2.1 Freundlich isotherms.

The non-linear isotherms are usually obtained in most sorption studies; the most common sorption isotherm for hydrophobic organic compounds is the Freundlich equation which considers multiple types of sorption sites (Delle Site 2001, Chiou 2002, Doucette 2003, Schwarzenbach et al. 2003, Xing et al. 2008).

Langmuir Isotherms - If sorption increases to a maximum value with C_{eq}, the data will often fit an equation of the form:

$$q = q_{max} \frac{K_L C_{eq}}{1 + K_L C_{eq}} \qquad (2.7)$$

The Langmuir isotherm describes the situation where the number of sorption sites is limited, so a maximum sorption was obtained. K_L is the "Langmuir" equilibrium constant and q_{max} is the maximum adsorbed amount, at which the monolayer coverage is formed. Note that if $K_L\, C_{eq} \ll 1$, the equation is linear. At low concentration a Langmuir isotherm may appear to be nearly linear. This model represents the maximum adsorption capacity (monolayer coverage) and constant binding energy between surface and adsorbate.

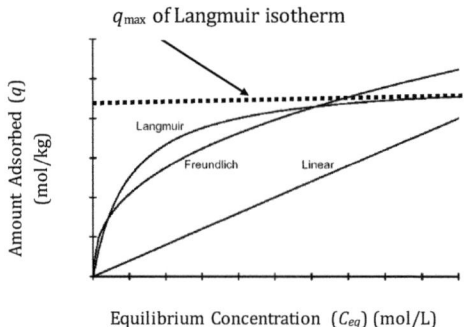

Figure 2.2 Sorption isotherms.

2.3 Sorption mechanisms

Most of the organic contaminants in the environment are hydrophobic. Thus, they have low water solubilities. The sorption on hydrophobic sites and sites with low polarity of OM is mainly based on entropy changes (solvent motivated sorption such as hydrophobic interaction) or relatively weak enthalpy forces (sorbent motivated sorption such as van-der-Waals and hydrogen bonding). Hydrophobic sorption is considered a companied effect of both mechanisms, while for more polar organic chemicals, sorption mechanisms can occur via different processes such as ionic exchange, charge transfer, ligand exchange and hydrogen bonding (von Oepen et al. 1991a, Kah 2007, and Keiluweit et al. 2009).

(a) Hydrophobic sorption

Hydrophobic sorption is considered the main sorption mechanism for hydrophobic organic chemicals through the sorption into the hydrophobic active sites of HSs. This type of sorption is a solvent motivated sorption and can be regarded as partitioning between a solvent and a non-specific surface. Hydrophobic partitioning processes are in general not affected by pH. While the dissociation of some HA functional groups at low pH can reduce the potential of HA for hydrophobic sorption, the creation of water-protected sites at pH < 5 can also create hydrophobic sorption sites of HS at low pH.

(b) Van-der-Waals interactions

Van-der-Waals forces includes three types of different interaction forces :

i. London forces (describing the dispersion interactions),
ii. Keesom forces (describing the dipole-dipole interactions),
iii. Debye forces (describing the interactions between dipoles and induced dipoles).

i Dispersion (London) forces

The sum of weak interactions between atoms, as well as between non-polar molecules has been shown by London (1930). It is considered that due to instantaneous electron distribution in a single molecule, a considerable dipole (charge separation) would be observed in which a small, and non-permanent, instantaneous distribution of charge would occur. Variation of charge distribution, which has an electric field (ϵ) associated with it, will induce a dipole in its neighbors.

ii Dipole (Keesom) forces

These forces are the first moment of the charge distribution in a molecule, which is termed the dipole moment and is measured in units of Debye (D). A dipole moment of one Debye is defined as the moment which results from the separation of two point charges equal in magnitude to the charge of an electron by a distance of 0.208 A°. Dipole-dipole forces depend on their mutual orientation, and thus exhibit a strong angular dependence and in solution, dipoles are not fixed and freely rotate in space. A very strong dependence of the interaction energy on the dipole moment was found, and in a pure dipolar liquid, the dependence is with the fourth power. In this type of forces, the molecules interact more through their individual bond or functional group dipoles than by their net dipole moment.

iii Dipole-induced dipole (Debye) forces

Debye has investigated intermolecular interactions through the induction of a dipole in one molecule by the permanent dipole of another molecule. The energy of these forces (induction energy) is produced via the interaction between the permanent electric field of one molecule (its dipole or quadrupole) with the polarizability of a second molecule. This type of interaction is weaker than the dipole-dipole forces.

(c) Hydrogen bond interactions

Hydrogen bonding is a specific interaction. It is highly dependent on the specific atoms involved and on the orientation of the molecules participating in the interaction. It occurs when a hydrogen atom is covalently bonded to an electronegative element (principally fluorine, oxygen, or nitrogen) and simultaneously interacts with lone electrons on a nearby electronegative element (or in some cases the π system of aromatic rings). It is also a very short-range interaction, it depends on dispersion and dipole interactions and it also involves a reasonable degree of electron pair sharing, i.e. covalency. Hydrogen bonding is highly probable for ionizable compounds although there is a strong competition with water molecules. The empirical solvatochromic technique is used to measure both hydrogen bond acidity and basicity. Their values can be obtained and used for interpreting the retention of solutes capable of accepting or donating hydrogen bonds (Vitha et al. 2006).

(d) Charge transfer

HSs have both electron-deficient and electron-rich regions, which results in charge-transfer formation via electron donor-acceptor mechanisms (π-π transfer). This is more probable for ionizable organic compounds.

2.4 Factors affecting sorption coefficient

Many factors can potentially affect the distribution of an organic chemical between an aqueous and a solid phase. These include environmental variables such as pH, temperature, ionic strength, dissolved OM, and co-solvents. In addition, there are factors related specifically to the experimental determination of sorption coefficients, such as sorbent and solid concentrations, equilibration time, and phase separation technique (Doucette 2000 and Delle Site 2001).

(a) Salinity or ionic strength

Salts can affect sorption of organic compounds by displacing cations from the soil matrix, by changing the activity of the sorbate in solution, and by changing the charge density associated with the soil sorption surface. Salt effects are most important for basic sorbates in the cation state, where an increase in salinity can significantly lower the sorption coefficient, while they are less important for neutral compounds, which may show either an increase or a decrease in sorption as salinity increases.

(b) Co-solvents

Miscible organic solvents, such as methanol and ethanol, have been shown to increase solubility of hydrophobic organic solutes and decrease sorption (Delle Site 2001). This effect is the result of (i) reducing the activity coefficient of the sorbate chemical in the aqueous phase, and (ii) competition for sorbing sites.

(c) Loss of compound

It has been found that in the sorption process, a loss of compound by sorption onto the walls of the equilibration vessels, volatilization, and chemical or biological degradation can affect the experimental determination of sorption coefficients. It is preferred to measure the chemical concentrations in both phases and determine the mass balance.

(d) Dissolved organic matter (DOM)

Differences in the DOM nature relevant to the origin and type and the environmental characteristics will lead to different molecular interactions with the sorbate, and to a variation of sorption affinity. It plays an important role in the transport and fate of the contaminants in the environment. In addition, it decreases the bioavailability of chemicals in solution and their toxicity.

(e) Sorptive concentration

Despite heterogeneity in the chemical composition of sorbents (solids), sorptive concentration has perhaps the most significant influence on the accumulation of a sorbate on or within a sorbent. At low sorptive concentration, the principle of chemical equilibrium predicts that increasing (or decreasing) sorptive concentration will result in an increase (or decrease) in sorbate concentration.

$$K_d = \frac{\text{Concentration in the solid phase } C_s}{\text{Concentration in the aqueous phase } C_w} \tag{2.8}$$

- At $C_s > C_w$ sorption is enhanced by decreasing the solute concentration.
- At $C_s < C_w$ sorption is reduced by increasing the solute concentration.

(f) pH

HA contains ionizable groups such as -COOH, -OH, and -NH_2, their polarization depends on pH of the medium. Therefore, sorption dependence on pH is predictable. However, for ionizable organic chemicals, sorption coefficients can be greatly affected, since pH affects not only the speciation but also the surface characteristics of natural sorbents.

(g) Temperature

Temperature is an important parameter that can influence the equilibria and rates of environmental processes. It can control the sorption equilibrium and the sorption kinetics for organic micropollutants. For most compounds, equilibrium sorption decreases with increasing temperature. However, different effects can be found, first an increase in equilibrium sorption with increasing temperature, which may be attributed to an increase in the rate of fast desorption with increasing temperature. Second, in some cases no effect of temperature on sorption equilibrium is found.

2.5 Experimental and estimation methods for determination of sorption coefficient

2.5.1 Experimental methods

(a) Batch equilibrium technique

A known amount of (sorbent) soil is dissolved in a known volume of water and placed in a vial and spiked with certain concentrations of solute solution in water (distilled water). The adjustment of the ionic strength of the above mixture is carried out by addition of milligrams of an electrolyte such as NaCl or $CaCl_2$. A minimum of headspace is left to avoid loss of solute concentration by volatilization to the vapor phase. The vials are shaken for a suitable time to reach the equilibrium, and then there are many analytical tools that can be used. The solute concentration will be divided by a certain ratio between two different phases in contact, the aqueous phase and the solid phase. The concentration of solute can be determined by measuring the difference between the initial concentration of the solute and the concentration after equilibrium has been reached (von Oepen et al. 1991a, Doucette 2000, Krop et al. 2001, OECD Guideline for Testing of Chemicals 106 2002, Endo 2008).

(b) Headspace technique

The headspace technique is used for the determination of sorption coefficients from vapor phase in the absence of or with a limited aqueous phase. The system consists of a solid phase sorbent directly in contact with the gas phase in sealed vials. Alternatively, a very limited aqueous phase is spiked with solute concentration and the molecules will enter the vapor phase. In equilibrium, the molecules in vapor phase equal those in solid phase or aqueous phase. After reaching equilibrium at a certain temperature, a volume of headspace vapor is withdrawn from the vials and analyzed. This technique is essential for very volatile compounds with high Henry's law constants (Paschke et al. 1999, Kopinke et al. 1999, and ter Laak et al. 2009).

(c) Fluorescence quenching

Fluorescence quenching is used for the determination of sorption constants for the association of PAHs with dissolved HAs and FAs and to study the sorption kinetics. The method is based on the fact that fluorescence intensity is proportional to the free concentration of PAHs in the solution. The fluorescence intensities of PAHs are measured in the presence and absence of humic material, and then the steps are repeated at different concentration of HAs. From the linear relationship, the sorption coefficient can be calculated (Doucette 2000, Delle Site 2001, and Krop et al. 2001).

(d) Equilibrium dialysis

Equilibrium dialysis is used for measuring the association of organic chemicals with dissolved OM. The spiked compound solution is added to the water sample which is specified by certain pH, ionic strength, temperature etc. in a glass bottle. The dialysis cell is filled with a known concentration of HSs and then transferred to the bottle. The bottle is shaken at fixed temperature for the time necessary to reach equilibrium. The filtrate is removed from the dialysis cell and then the bottle is analyzed. The compound inside the dialysis cell consists of two fractions; one is free and dissolved while the other is bound to HSs. The bound concentration to HSs can be calculated from the difference between the concentration of the compound inside and outside of the dialysis cell (Doucette 2000, Delle Site 2001, Krop et al. 2001, and Mackenzie et al. 2002).

(e) Humic acid HPLC Column

This method focuses on the use of HA-HPLC column for the direct or indirect determination of log K_{oc} for organic compounds by measuring the retention times for the various compounds through the column using a series of mixtures of water and methanol (Delle Site 2001, Jonassen 2003, and Tülp et al. 2008, 2009). This method can be used to calculate the distribution coefficient of the compound directly, by determination of column coefficients

$$K_x = \frac{t_R - t_o}{t_o} \tag{2.9}$$

for both blank column $K_{x,\,blank}$, and soil organic matter column $K_{x,\,SOM}$. Log K_{oc} can be calculated using this equation proposed by Jonssen (2003).

$$\log K_{oc} = \log \frac{K_{x,SOM} - K_{x,blank}}{V_s \cdot \rho \cdot OC_{SOM}} V_m \tag{2.10}$$

t_R is the retention time of the solute,
t_0 is the retention time of methanol,
OC_{SOM} is the organic carbon content from SOM in column material,
V_m is the volume of the mobile phase in the HPLC column,
$V_S = (V_{total} - V_m)$ is the volume of the stationary phase,
V_{total} is the total inner volume of the HPLC column,
ρ is the density of the stationary phase.

2.5.2 Estimation methods

(a) Estimation from hydrophobicity factor (log K_{ow})

Many authors have demonstrated the estimation of sorption coefficients (log K_{oc}) from the octanol-water partition coefficient (K_{ow}), as octanol has properties similar to the organic phase (Karickhoff 1981),

$$\log K_{oc} = a \log K_{ow} + b \tag{2.11}$$

Chin et al. (1989) have found that the partition constants of organic compounds to HA (K_{oc}) are 0.2-1.5 log units lower than their K_{ow} values and that these differences increase with increasing hydrophobicity. However, this method is limited and only suitable for neutral compounds. It is not suitable for the prediction of the sorption behaviour of ionogenic compounds.

(b) Molecular descriptors and QSAR

Leo et al. (1971) have proposed that the thermodynamic parameters of organic compounds can be used for prediction methods based on the fact that a molecule is a collection of molecular fragments. Each fragment can contribute to the thermodynamic properties.

Sabijić (1987) has proposed a prediction method depending on the molecular topology and quantitative structure-activity relationship (QSAR). The aim of this method is to find a relationship between the molecular structure of the organic compound and its properties. For sorption processes, the parameters of relationship between the sorption of chemicals by soil and sediments and their structural parameters such as the molecular connectivity indices (MCIs) have been investigated, and they have been successfully used in the estimation of environmental constants (partition coefficients).

(c) LSER approach

The linear solvation energy relationships (LSER) developed by Abraham et al. (1987), (1991), (1994), (1999) and (2004) consider contributions towards free energy changes from multiple kinds of molecular interactions with both water and bulk organic phases. This LSER model relies on two linear free energy relationships, one for solute transfer between two condensed phases and the other for processes involving a gas in condensed phase transfer. It is used for the prediction of the partitioning equilibrium due to its ability to split the intermolecular interactions into many types of forces (such as dispersion forces, dipole-dipole interaction, dipole-induced dipole interaction and H-bonding interactions). Directly interpretable solute descriptors are obtained, and by using multiple regressions, the phase descriptors can be generated (Abraham et al. 1987, 1991, 1994, 1999 and 2004).

The solute transfer between two condensed phases can be expressed as the following equation:

$$\text{Log Property} = c + e \cdot E + s \cdot S + a \cdot A + b \cdot B + v \cdot V \qquad (2.12)$$

$$\text{Log}(K_{oc}) = c + e \cdot E + s \cdot S + a \cdot A + b \cdot B + v \cdot V \qquad (2.13)$$

E, S, A, B and V represent the compound descriptors for excess molar refraction (E) in cm^3 mol^{-1}, dipolarity / polarisability (S), hydrogen bond acidity (A), hydrogen bond basicity (B) and McGowan volume of solute (V) in cm^3 mol^{-1}. In case of a model that relies on processes involving a gas in condensed phase transfer depending on the solute but independent on the solvent, the logarithm of the solute gas phase dimensionless Ostwald partition coefficient in hexadecane at 289 K (L parameter) is used instead of V. V and L parameters are measured for the endoergic effect of disrupting solvent-solvent interactions. The coefficients e (solvent dispersion interaction), s (the ability of solvent phase to undergo dipole-dipole and dipole-induced dipole interactions with a solute), a (the complementary solvent hydrogen bond basicity), b (the solvent phase hydrogen bond acidity) and v (endoergic cavity effect + exogeric solute-solvent effect) are calculated from solutes descriptors via multiparameter linear regression analysis of the data set. These coefficients depend on the solvent but are independent of the solutes, and the constant c (solvent specific free energy contribution) depends on volume entropy effects. This approach is considered one of the most useful approaches for the analysis and prediction of sorption coefficients, though it is not applicable for ionizable solutes. The recent form of the LSER approach (Abraham equation) can be used to predict compound properties such as log K_{ow}, or log K_{oc} for new untested compound sets depending on their solute specific Abraham parameters (E, S, A, b and V) which are available experimentally or estimated in the literature. We should keep in mind that the applicability domain of the resulting LSER model is directly determined by the diversity of the used calibration dataset. Recently some proposed LSER models have been published to explain and describe the intermolecular interactions in sorption process to OM (Endo et al. 2009a, Bronner et al. 2010, Neale et al. 2012).

2.6 Nernst's Distribution Law

According to Nernst's Distribution law (1891) or Partition law, "When a solute is taken up with two immiscible liquids, in both of which the solute is soluble, the solute distributes itself between the two liquids in such a way that the ratio of its concentration in the two liquid phases is constant at a given temperature provided the molecular state of the distributed solute is same in both the phases".

$$K_d = \frac{C_1}{C_2} \qquad (2.14)$$

Where C_1 and C_2 are the concentrations of the solute in two phases. K_d is called distribution coefficient or partition coefficient. The Nernst's distribution law permits us to determine the most favourable conditions for the extraction of substances from solutions.

When solute undergoes association in one of the solvents, we have the following form,

$$K_d = \frac{C_1}{n\sqrt{C_2}} \quad \text{or} \quad \frac{n\sqrt{C_1}}{C_2} \tag{2.15}$$

where n =order of association.

When solute undergoes dissociation, we have the following form,

$$K_d = \frac{C_1}{C_2(1-\alpha)} \quad \text{or} \quad \frac{C_1(1-\alpha)}{C_2} \tag{2.16}$$

where α= degree of dissociation

2.7 Equilibrium partitioning of pollutants in the environment

Micropollutants in the environment can be distributed among the various environmental compartments (e.g. air, water, soil, vegetation) as a result of various transport processes. The net transport of pollutants from one compartment to another is limited by equilibrium constraints which are quantified by appropriate partition coefficients. The partition coefficient K_{ij} is defined as the ratio of the equilibrium concentration of a pollutant in one environmental compartment with respect to another environmental compartment,

$$K_{ij} = \frac{C_i}{C_j} \tag{2.17}$$

where C_i is the concentration in compartment i and C_j is the concentration in compartment j. K_{ij} can be determined experimentally in laboratories, and it can also be estimated from theoretical and empirical relations. To understand the partitioning process, the concept of the equality of chemical potential (or fugacity) for a given chemical present in a multiphase system which is at equilibrium should be known. Figure 2.3 shows a simple multicompartment environmental system which consists of air, water and soil compartments.

In a simple environmental system, the water compartment can consist of water and suspended solids, and the soil compartment consists of soil-air, soil-solids, soil-water and possibly a soil organic free phase. At equilibrium, the fugacity of a chemical which is introduced to the above closed system is the same in the four phases (Chiou 2002, and Schwarzenbach et al. 2003).

$$f_{air} = f_{water} = f_{soil\ matrix} = f_{sediment} = f_{suspended\ solids} \qquad (2.18)$$

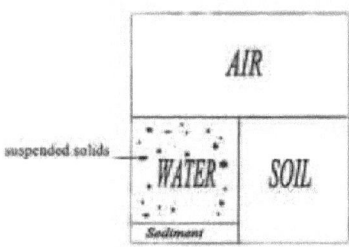

Figure 2.3 Simple multiphase environmental systems.

2.8 pH Effect on equilibrium partitioning

When a change in environmental conditions (e.g. pH variation) occurs, dissociation and protonation can occur in the aqueous phase. In the case of acids (AH) and bases (B), donating or accepting a proton is possible. According to the Henderson-Hasselbalch relationship, the ionized fraction, f_i, and unionized fraction, f_u, can be calculated depending on the values of both pH and pK_a (Kah et al. 2007, Franco et al 2008 and 2009).

Ionisation:

The dissociation of a weakly acidic compound can be described as the following equation:

$$AH + H_2O \rightleftharpoons A^- + H_3O^+ \qquad (2.19)$$

$$C_T \rightleftharpoons A^- + AH \qquad (2.20)$$

where H_3O^+, A^-, HA and C_T are defined as the aqueous concentration of hydronium ion, anionic, neutral species, and total concentration of solute respectively.

The dissociation constant can be expressed as the following form:

$$K = \frac{A^-.H_3O^+}{AH.H_2O} \rightarrow K_a = \frac{A^-.H^+}{AH} \rightarrow A^- = AH\frac{K_a}{H^+} \qquad (2.21)$$

By substitution in equation (2.20), the total concentration can described as the following equation:

$$C_T \rightleftharpoons A^- + AH = AH\left(1 + \frac{K_a}{H^+}\right) \qquad (2.22)$$

$$\frac{AH}{AH+A^-} = f_u = \frac{1+\frac{K_a}{H^+}AH}{AH\left(1+\frac{K_a}{H^+}\right)} = \frac{1}{1+\frac{K_a}{H^+}} \qquad (2.23)$$

The unionized fraction f_u can be calculated from the equation (2.23), and its form can be modified so it provides the ratio of the neutral species to the anion as a function of pH as shown in equation (2.24) and also in Figure 2.4.

$$f_u(\text{acid}) = \frac{1}{1+10^{pH-pK_a}} = f_i(\text{base}) \qquad (2.24)$$

Calculation of dissociation

For given pH and pK_a values, the ionized and unionized fractions f_i and f_u can be calculated for mono-acid and mono-base according to the following equations (Zhao et al. 2010).

$$f_{u\,(\text{acid})} = f_{i\,(\text{base})} = \left[\frac{1}{1+10^{pH-pK_a}}\right] \qquad (2.25)$$

$$f_{i\,(\text{acid})} = f_{u\,(\text{base})} = \left[\frac{1}{1+10^{pK_a-pH}}\right] \qquad (2.26)$$

while equation (2.27) holds for di-acid

$$f_{u\,(\text{acid})} = f_{i\,(\text{base})} = 1 - f_{u\,(\text{base})} = \left[\frac{1}{1+10^{pH-pK_{a1}}+10^{pH-pK_{a1}-pK_{a2}}}\right] \qquad (2.27)$$

The octanol–water partition coefficient of the neutral molecule ($K_{ow(unionized)}$) and of the ion ($K_{ow(ionized)}$) can be calculated at the overall given pH. Prediction model for hydrophobicity of acids depending on the Henderson-Hasselbalch equation was proposed by Kah et al. (2008).

$$\text{Log P} = \log\left[10^{(\log P_n + 10^{\wedge}(\log P_i + pH - pK_a))}\right] - \log\left[1 + 10^{(pH - pK_a)}\right] \quad (2.28)$$

Log P_n, log P_i the hydrophobicity of the neutral and anionic species.

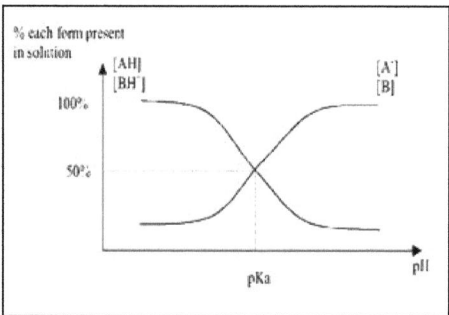

Figure 2.4 Presence of neutral or ionic form in solution according to the pH and pK_a. [AH] and [BH+] are the protonated forms, [A-] and [B] the dissociated forms of the acidic and basic compounds respectively (Kah et al. 2007).

2.9 Temperature effect on equilibrium partitioning

The partition constant $K_{w,s}$ related to temperature can be described by the following equation (Atkins 1994 and Schwarzenbach et al. 2003).

$$K_{w,s} = e^{-\Delta_{ab}G/RT} \quad (2.29)$$

G is the free energy difference between water and soil constant of compound (x) at two different temperatures T_1, T_2, T is the absolute temperature in K, and R is the universal gas constant.

Other formula for the above equation:

$$\ln K_{x(w,s)} = -\frac{1}{R}\left(\frac{\Delta_{(w,s)}Gi}{T}\right) \quad (2.30)$$

By differentiation with respect to temperature,

$$d \ln K_{x(w,s)} = -\frac{1}{R}d\left(\frac{\Delta_{(w,s)}Gx}{dT}\right) \quad (2.31)$$

By applying the Gibbs-Helmholtz equation (Atkins 1994):

$$\text{Therefore, } \frac{1}{R} \cdot \frac{\Delta_{(w,s)}Gx K}{dT} = \frac{1}{R} \cdot \frac{\Delta_{(w,s)}Hx}{T^2} \quad (2.32)$$

Which leads to the van't Hoff equation:

$$\frac{d \ln K_{x(w,s)}}{dt} = \frac{\Delta_{(x,t)}Hx}{RT^2} \quad (2.33)$$

$\Delta_{(w,s)}H_x$ is the standard enthalpy change term of partitioning of compound (x) between two different temperatures T_1 and T_2. By integration of equation (2.33) of enthalpy change:

$$\frac{\ln K_{x(1,2)}(T_2)}{K_{x(1,2)}(T_1)} = \frac{\Delta_{(1,2)}Hx}{R} = \left(\frac{1}{T_2} - \frac{1}{T_1}\right) \quad (2.34)$$

$$K_{x(1,2)}(T_2) = K_{x(1,2)}(T_1)e^{\frac{-\Delta_{(1,2)}Hx}{R}\left(\frac{1}{T_2} - \frac{1}{T_1}\right)} \quad (2.35)$$

By measuring the partition constant at various temperatures, ΔH and ΔS can be obtained by a linear regression.

$$\frac{\ln K_{x2}}{K_x} = \frac{\Delta Hx}{R}\left(\frac{1}{T_1} - \frac{1}{T_2}\right) \quad (2.36)$$

K_{x1} is the equilibrium constant at absolute temperature T_1 and K_{x2} is the equilibrium constant at absolute temperature T_2

$$\text{Since } \Delta_{(1,2)}G_x = \Delta_{(1,2)}Hx - T\Delta_{(1,2)}S_x \quad (2.37)$$

$$\Delta_{(1,2)}G_x = -RT \ln K_{x(1,2)} \quad (2.38)$$

Then

$$\ln K_x = \frac{\Delta H_x}{RT} + \frac{\Delta S_x}{R} \qquad (2.39)$$

The slope of the line is equal to $-\Delta H/R$, and the intercept is equal to $\Delta S/R$.

2.10 Humic substances

HSs consist of a series of highly acidic, high molecular weight polydispersed polyelectrolytes with high degree of molecular irregularity and heterogeneity. They have been considered to represent 60 - 70 % of SOM. HSs are characterized by a high content of oxygen, as they possess a variety of functional groups containing oxygen, such as COOH, phenolic, aliphatic and enolic OH and C=O, amino, heterocyclic amino, imino, and sulfhydryl groups (Stevenson 1976).

HA is less acidic than FA due to lower content of COOH groups. No general structural formula for HA can be given that would account for the theoretical structure of HA. Choudhry (1984) has shown the presence of aromatic structures, carbohydrates, aliphatic chains, free radicals, peptide linkages, and chinoide structures in HA structure.

As described in Jonassess (2003) the organic carbon content of aldrich HA in the literature is between 8-13% wt. to 69% wt. Infrared spectroscopy showed that the backbone of aldrich HA consists of both aromatic and aliphatic parts, and contains both carbonyl groups and carboxylic acids. Aldrich HA, which is well characterized in the literature, consists of smaller molecules (approximately 4500 Da), with a higher content of aromatic carbon. The presence of aromatic structures was detected by GC-MSD data and the presence of some identified masses of PAHs such as specifically naphthalene (m/z 128).

Choudhry (1984) has shown that HA contains both hydrophilic and hydrophobic sites which are responsible for the reduction of solvent sorption, carboxylic and phenolic groups. It is very important to mention of the protonation state of HA. Regarding the HA structure change, 72% of the H-C groups of humic fraction exist as $(CH_2)_n$-CH_3 group. The long chain aliphatic structure present in HSs may contribute in the sorption of environmental solutes. HA has some region of polymeric three dimensional cross linked structure and its configuration can hinder the sorption process due to the restricted motion of crystalline or glassy state (site limitation of sorption can occur).

The HA structure is affected by the change of pH. According to the Henderson-Hasselbalch equation (Doucette 2003, Kah et al 2007, Franco et al. 2008, and 2009), it is known that the dissociation constant of HA is in the range of 3 - 6. This means that a neutral form of HA will be available at pH 10 and 7. However, at a pH value of the solution equal to the pK_a of HA, 50 % of both neutral and anion forms will be available together. Therefore, it will be expected that at pH 4, a slight deprotonation of carboxylic groups would occur, as well as the ionization of other functional groups of HA, of OH and NH_2, in addition to the protonation and deprotonation of ionizable compounds in that range of pH such as the following functional groups (calculated by ACD/pK_a v.12 software):

- aliphatic COOH: in alkyl chain pK_a ca. from 4.8 to 5.0,
- aromatic COOH: pK_a ca. 4.2,
- carbonyl group C=O (aliphatic and aromatic chain): no dissociation or protonation,
- Ph-OH aromatic chain: pK_a ca. 10.0,
- alkylphenols: pK_a ca. (10 ... 11.2),

pK_a of conjugated acids of bases as follows:

- aliphatic primary amino group -NH_2 : ca. (9.9 ... 10.9),
- aliphatic secondary amino group -NH- : ca. (10.7 ... 11.2),
- aromatic primary amino group -NH_2 : ca. (3.8 ... 5), and
- aromatic secondary amino group -NH- : ca. (4.4 ... 7.3).

2.11 Humic substances role in soil photolysis

Some phototransformation is occurring in surface waters through light absorption by dissolved OM in the range of 300-500 nm, which act as sensitizers or precursors for the production of reactive intermediates. The reactive species can be produced by irradiation of HAs and FAs extracted from different soils or aquatic media. These reactive species can lead to the photodegradation of many organic pollutants under the action of sunlight in the presence of HSs, they act as photoinductors (or photosensitizers) to produce the excited intermediates. The main species of HSs following irradiation is the excited triplet state which is responsible for the photodegradation of aquatic pollutants in surface waters through direct reactions. HSs can react with O_2 via energy or electron transfer process to create highly reactive species such as $^1O_2$, $OH^·$, superoxide anion ($\overline{O_2}$), and peroxide radical $OOH^·$ (Boule et al. 1999 and Katagi 2004).

Excited triplet states of HSs are the main species responsible for the photoinduced degradation of aquatic pollutants. It is also possible for the excited triplet state of HSs to transfer their energy to the ground state of organic molecules, but only if the energy level of the triplet state of the organic molecules is lower than that of HSs. This photoprocess can be described as energy, electron or hydrogen atom transfer reactions, it was described by equation (2.40).

$$^1HS \xrightarrow{h\upsilon} {}^1HS^* \xrightarrow{ISC} {}^3HS^* \xrightarrow{^3O_2} {}^1HS + {}^1O_2 \xrightarrow{RH} Products \qquad (2.40)$$

^1HS	The singlet ground state of HSs
$^1HS^*$	The singlet excited state of HSs
$^3HS^*$	The triplet excited state of HSs
$^1O_2$	The singlet ground state of oxygen
$^3O_2$	The triplet excited state of oxygen
$h\upsilon$	Quantum energy of photon, h = Planck's constant = 6.626 x 10^{-34} Js, υ is frequency of radiation
ISC	Inter-system crossing
RH	Reduction process

It is also possible that singlet state oxygen will be formed as result of energy transfer between the triplet state of HSs and the triplet state of molecular oxygen. The quantum yield of singlet state oxygen formation depends on the irradiation wavelength and the natural HSs. When irradiation is performed at 254 nm, the rate of photodegradation decreases due to the competitive light absorption by colored HSs, and then direct photolysis will be inhibited. The potodegradtion of organic compound by indirect method (HSs effect) has not investigated in the current study.

3. Materials and Methods

3.1 Materials

3.1.1 Organic chemicals

Sorbent: In this study, Aldrich HA has used in the form of sodium salt provided from Aldrich (Steinheim, Germany) as organic matter sorbent. Its structure is shown in Figure 3.1.

Sorbate: (organic compounds): Sixty one substances have been chosen according to their hydrophobicitiy, volatility, structure, water solubility and possibility of precisely detection by chromatographic analysis. The organic chemicals have been divided into fifteen that have an acidic character as phenols, seventeen that have a basic character as triazines, and twenty nine that have a neutral character. These sixty one organic substances have been categorized to seven various chemical classes regarding to their chemical structures [three anilines, nine benzene and substituted benzenes, three biphenyls, twelve heterocyclic, fourteen phenols, twelve PAHs, and eight organochlorines]. These organic chemicals that have been used as organic sorbates were highly standard grade and have been provided by different suppliers. Their CAS Numbers, molecular weights, smiles and providers are shown in Appendix 9.2. The purity of the investigated organic substances has been ≥ 97%. The chemical structures of the probe sorbates are illustrated in Scheme 3.1.

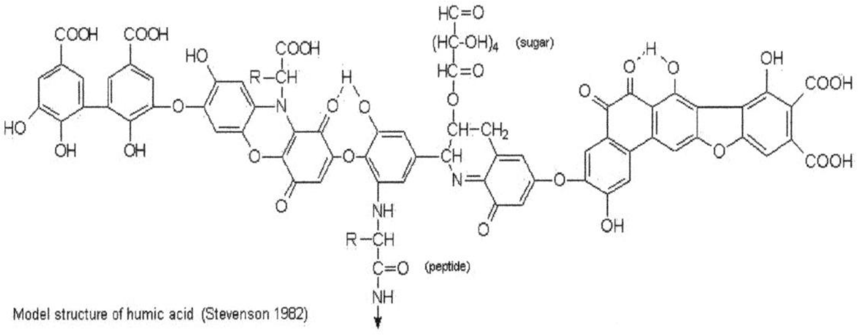

Figure 3.1 The model structure of HA (Stevenson 1982).

Solvents:

2-Propanol, cyclohexane, acetonitrile, and methanol have been provided by Merck KGaA (Darmstadt, Germany).

Reagents:

The different reagents have been used for the preparation of HA in different buffer solutions at different $pH^{(25\ °C\pm2)}$(4-10±0.2) as di-sodium hydrogen phosphate, potassium hydrogen phthalate, 1M hydrochloric acid, 2M sodium hydroxide. Sodium azide 200 mgL^{-1} (Ilani et al. 2005) has been used as an antimicrobial agent. These have been provided by Merck (Darmstadt, Germany).

3.1.2 Equipments

SPME fibre assembly:

Different types of fibres have been used according to their coating materials. Polydimethylsiloxane (100 μm PDMS) for PAHs and organochlorines, Polydimethylsiloxane – Divinylbenzene (65 μm PDMS-DVB) for amino and nitro PAHs, and Polyacrylate (85 μm PA) for phenols and triazines, of manual and automated holders with the 23 gauge have been used for volatile-semi volatile substances as shown in Figure 3.2. 100 μm PDMS and 65 μm PDMS-DVB have been thermally desorbed at 250 °C for half an hour while 85 μm PA has been conditioned at 280 °C for one hour to be used as an active phase in the adsorption of gaseous form of organic substances. Fibres have been provided from Supelco (Bellefonte, PA, USA) as shown in Scheme 3.2.

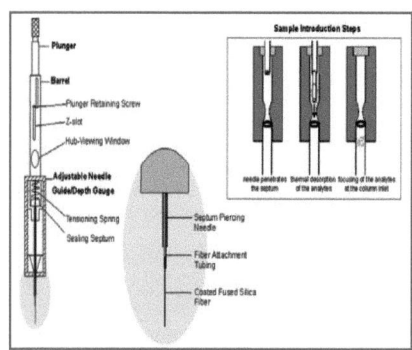

Scheme 3.2 SPME fibre assembly (Pawliszyn 2001).

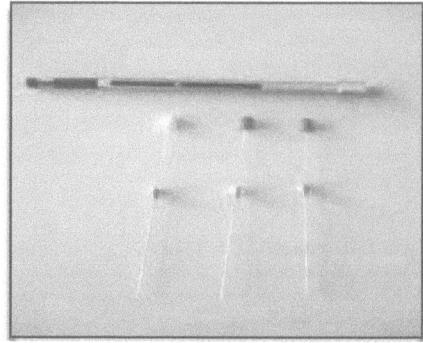

Figure 3.2 SPME different fibres. Red-100 μm PDMS, white- 85 μm PA, blue- 65 μm PDMS-DVB.

	1	2	3	4
A	Aniline	4-Chloroaniline	2,3,5,6-Tetrachloraniline	Benzene
B	Benzaldahyde	Toluene	Fluorobenzene	1,2,4-Trimethylbenzene
C	1,3,5-Trichlorobenzene	1,2,3,4-Tetrachlorobenzene	Pentachlorobenzene	Hexachlorobenzene
D	2-Fluorobiphenyl	PCB 180	PCB 202	Aldrin

	1	2	3	4
E	Lindane	2,4`-DDD	2,4`-DDE	2,4`-DDT
F	4,4`-DDD	4,4`-DDE	4,4`-DDT	Naphthalene
G	1-Nitronaphthalene	2,3-Dimethylnaphthalene	Anthracene	2-Aminoanthracene
H	2-Methylanthracene	Fluorene	2-Nitrofluorene	1-Aminopyrene

	1	2	3	4
I	Fluoranthene	Benzo(a)pyrene	6-Aminochrysene	Phenol
J	4-Chlorophenol	4-Nitrophenol	2,4-Dichlorophenol	2,4,6-Trichlorophenol
K	2,3,4,6-Tetrachlorophenol	Pentachlorophenol	2,4-Di-t-butylphenol	3-Phenylphenol
L	4,4´-Isopropylidenediphenol	4-n-Hexylphenol	4-n-Nonylphenol	4-n-Octylphenol

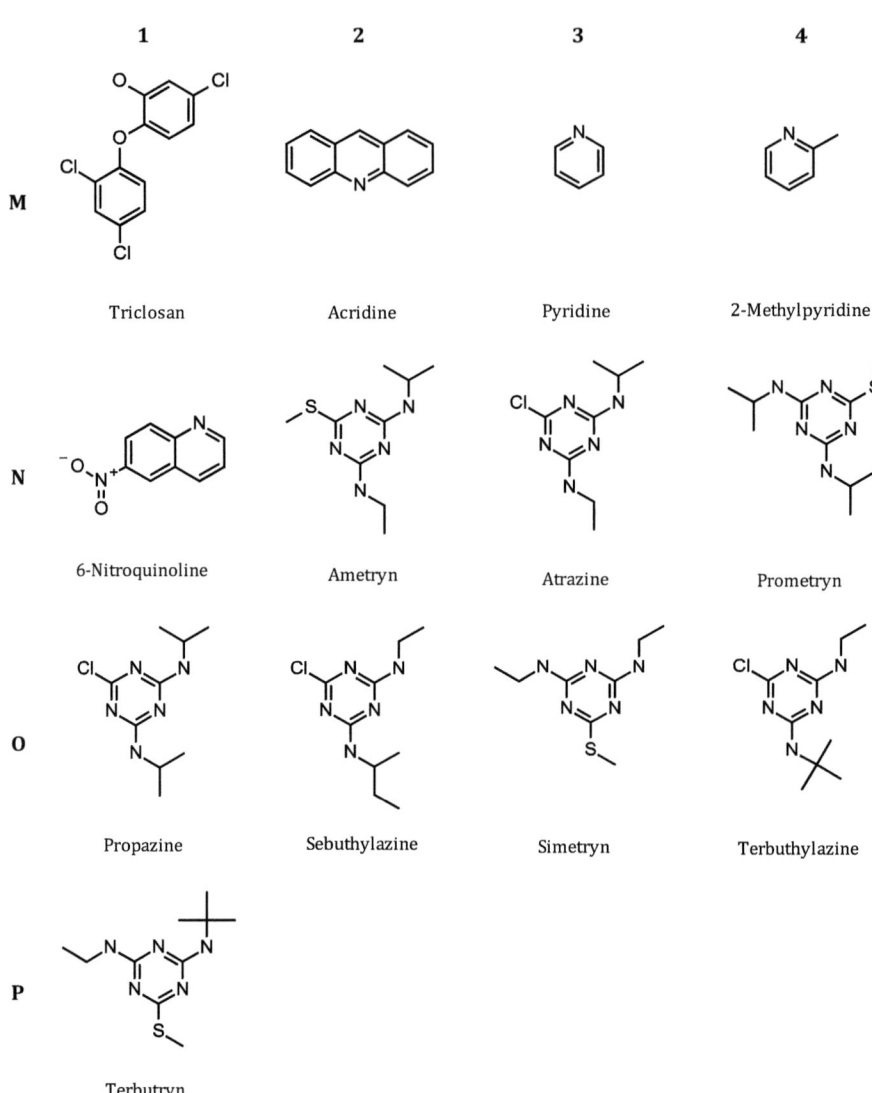

Scheme 3.1 The chemical structure of probe compounds.

pH:
The pH of the different solutions was measured by a Schott pH-meter type CG842/14PH (Schott, Germany) after its calibration with standard buffers at 25 ± 2 °C.

Electrical conductivity of the solutions:
The electrical conductivity of the different buffer solutions were measured using a WTW Microprocessor conductivity-meter LF-196 (WTW, Germany) and it has found 11.33±2 mScm^{-1}.

Ultrasonication:
Elma Ultrasonic type T570/H (Elma, Germany) has produced the ultrasonic vibrations to speed up the velocity of the insoluble substances solubilises to obtain a homogenous standard solution of organic solutes.

Stirring:
Heidolph Stirrer type MR3000D (Heidolph, Germany) has been used for the homogeneity of the HA solutions as well as in manual and back extraction of photolyzed solutions at a speed of 600 rpm.

Agitation:
Heidolph Shaker type Promax 3020 (Heidolph, Germany) has been used to reach the equilibrium in the sorption experiments at different temperatures (5-45) ± 2 °C and at a speed of 157 rpm.

Incubation:
A Heraeus incubator type BK 6160 (Heraeus, Germany) has been used to stabilize the temperature of the solutions for the sorption experiments obtaining the equilibrium at different temperatures of (5-45) ±2 °C.

3.2 Experimental methods

3.2.1 SPME optimization
Sets of experiments have been carried out in the beginning of the study to optimize the best conditions for the sampling process.

a-Fibre type selection
The fibre coating should be selected based on the coating type (polar or non polar), coating volume (influences sensitivity) and coating thickness (influences extraction time). In this study the choice of the fibre depended on the stationary phase polarity of the fibre coating (Krutz et al. 2003 and ter Laak et al. 2005).

b-Fibre negligible-depletion test

It was necessary in the beginning of the work to check whether the extraction by fibre did disturb the freely dissolved concentration of the analyte in the measurements. The SPME fibre should not take up more than 5% (Vaes et al. 1997) of each analyte freely dissolved concentration in the sorption experiment i.e. fluoranthene depletion ≤ 5%. The absolute mass of analyte and adsorbed amount by fibre have been estimated from the calibration plot using different concentrations in 2-propanol and water phase as shown in Figures 3.3 and 3.4.

Depletion % is calculated as follows:

$$\text{Depletion \%} = \frac{m_f}{m_w}(100) \tag{3.1}$$

m_f is the adsorbed mass of analyte by fibre, while m_w the absolute mass of analyte in the aqueous volume. Table 3.1 shows an example of depletion calculation.

c-Extraction mode

Extraction mode selection has been based on the sample matrix composition analyte amount, analytes volatility based on their air-water partition coefficients, the analytes hydrophobicity based on their 1-octanol-water partition coefficients, and the temperature of extraction process. In this study the direct and head space modes have been used for both volatile and non-volatile organic compounds depending on the above mentioned conditions (Pawliszyn 2001, Krutz et al. 2003, ter Laak et al. 2005 and Vaes et al. 1997) as shown in Figure 3.5 a, and b.

d-Extraction time and conditions

Sets of trials have been carried out in the beginning of the measurements to obtain the best conditions to reach the needed extraction time, which is the time required for an analyte to be extracted by negligible depletion by the fibre. Simultaneously up-taking the whole amount of the freely dissolved concentration of the analytes from aqueous phase or gaseous phase to obtain a precisely, a quantified sharp peak in the chromatogram as shown Figure 3.5. In the present study, different time ranges between 10 and 60 min as an adsorption time have been enough to find for an extraction at a different temperatures (5-45) ± 2 °C and at different pH values. The desorption time of the fibre has been 2-3 min, all the required conditions are listed in Appendix 9.10.

e-Injection mode

The time for desorption of SPME fibre via a manual injection was two min splitless while the times for the injection via SPME autosampler have been one and three min splitless.

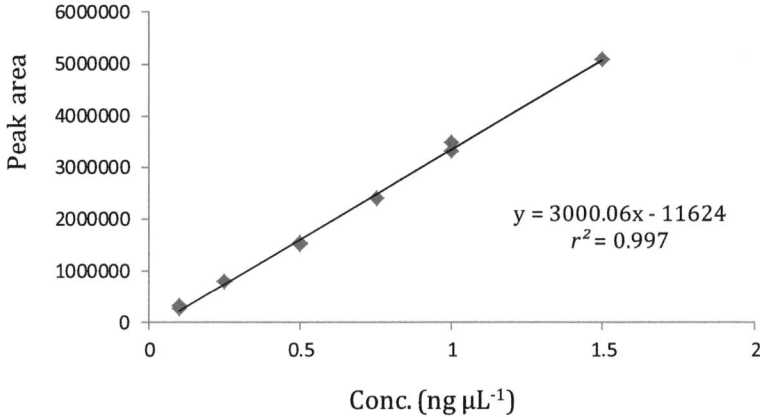

Figure 3.3 Calibration plot of fluoranthene in solvent (2-propanol) at 25 ±2 °C.

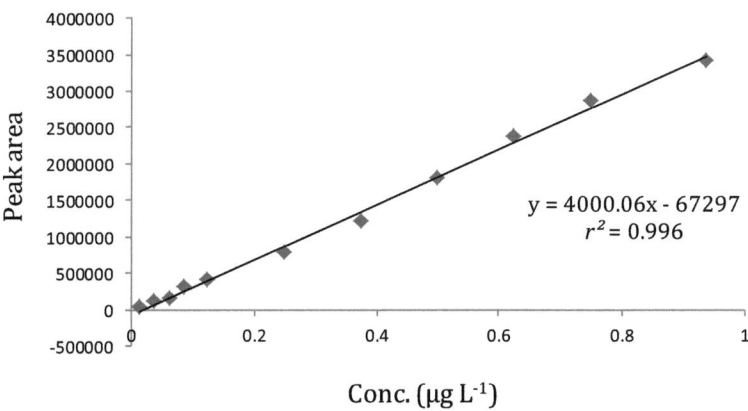

Figure 3.4 Calibration plot of fluoranthene in aqueous phase (water) in phosphate buffer pH7±0.2 at 25 ±2 °C.

Table 3.1 An example of depletion calculation for a fluoranthene probe in 20 mL vial.

Probe name	C_w (µg/L)	M_w (ng/Vial)	M_f (ng/Fibre)	Depletion %
pH 4 ref 4	0.75	15	0.421	2.8
pH 4 ref 5	1	20	0.618	3.1
pH 4 ref 6	1.25	25	0.772	3.1
pH 4 ref 7	1.5	30	0.985	3.3
pH4 ref 8	1.875	37.5	1.191	3.2
pH 7 ref 4	0.75	15	0.387	2.6
pH 7 ref 5	1	20	0.558	2.8
pH 7 ref 6	1.25	25	0.720	2.9
pH 7 ref 7	1.5	30	0.861	2.9
pH7 ref 8	1.875	37.5	1.021	2.7
pH 10 ref 4	0.75	15	0.317	2.1
pH 10 ref 5	1	20	0.377	1.9
pH 10 ref 6	1.25	25	0.480	1.9
pH 10 ref 7	1.5	30	0.597	2.0
pH 10 ref 8	1.875	37.5	0.686	1.8

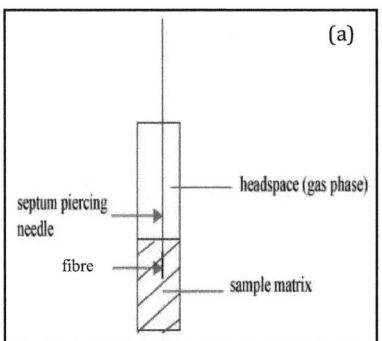
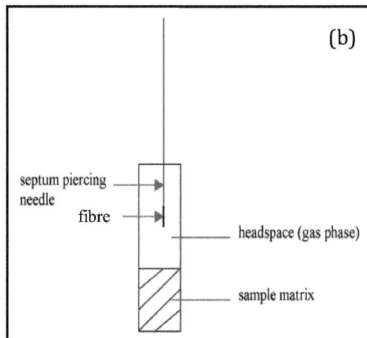

Figure 3.5 (a) Direct and (b) Headspace modes of SPME (Pawliszyn 2001, Krutz et al. 2003 and ter Laak et al. 2005).

f- Selection of the agitation technique

For aqueous samples, the agitation (needle vibration technique) is required in most cases to facilitate the mass transportation between the bulk of the aqueous sample and the fibre. A speed of 250 rpm was used for the agitator of the autosamplers.

g-Equilibration time

The choice of equilibrium time has been carried out with regards to the optimal measuring times, peak areas and the nature of the analytes and their 1-octanol-water partition coefficients. The equilibrium time has been optimized by determining the time required for an analyte to reach equilibrium between the solid phase (humic matrix) and the aqueous phase in the sorption experiments. Several experiments have been setup to determine the minimal time to reach the equilibrium as shown in Figure 3.6. The peak area has attained a stable value at such a time at which the equilibrium between the freely dissolved concentration of the analyte in aqueous phase and that of the bonded concentration in humic matrix have reached at a different incubator temperatures (5-45) $\pm$ 2 °C. For instance, fluoranthene, atrazine, hexachlorobenzene and 4, 4`DDT attained equilibrium after 24-48 h and were ready for analysis. In the present study, 24-48 h equilibration time was enough to reach a stable equilibrium state for all samples.

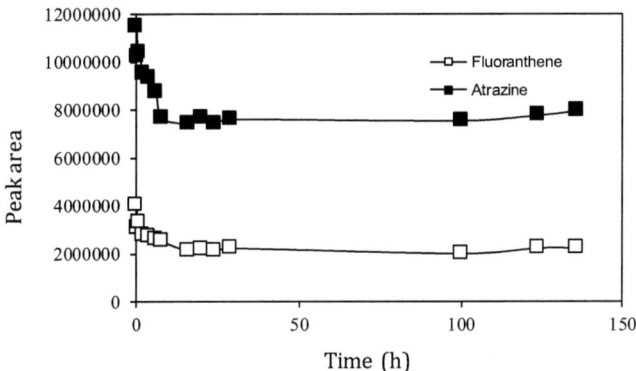

Figure 3.6 Equilibrium time plot of fluoranthene and atrazine in different pH values.

3.2.2 Check for saturation effect (isotherm measurement)

The isotherm curves for such compounds were obtained by plotting the bound concentration of the compound per humic matrix concentration $(C_{ref} - C_{humic}) \times \frac{V_w}{m_{humic}}$ (y axis) versus the freely dissolved concentration of the compound in the aqueous solution (x axis) (Xing et al. 2008). An example of the sorption isotherm of fluoranthene is shown in Figure 3.7. Saturation occurred when the straight line of linear fit becomes curved and stable after a certain critical concentration, for higher concentrations the calculated partition coefficients will be smaller due to the non-linearity.

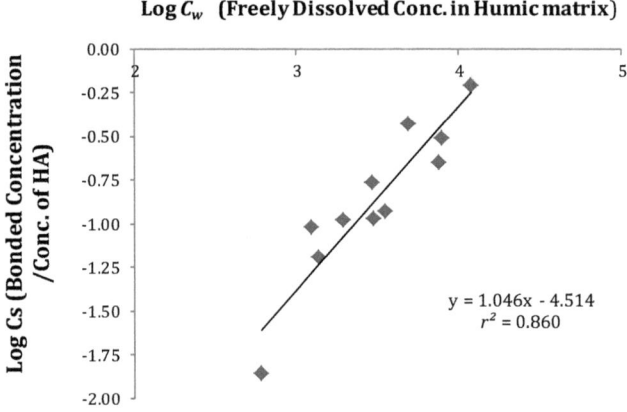

Figure 3.7 Sorption isotherm of fluoranthene, from the Freundlich equation (Xing et al. 2008). $Log\ C_s = log\ K_f + n.log\ C_w$, n is Freundlich exponent $\geq$ =1.046 it means sorption will be enhanced at higher concentrations of sorbate, K_f is Freundlich constant = $3.06 \cdot 10^{-5}$.

3.2.3 Batch method for sorption experiments
Elemental analysis of the humic acid

Elemental analysis and determination of total organic carbon fraction f_{oc} of HA have been carried out using the elemental analysis equipment as shown in Table 3.2. The total carbon, hydrogen, nitrogen and sulphur concentrations in the HA have been determined and the percentage of total organic carbon of Aldrich HA has been calculated using RC-412 (LECO Instruments Inc., Germany). A solid sample of the HA sodium salt was burned gradually by the thermal program. The total organic carbon program has started from 200 °C until 530 °C (2 min) and the peak temperature was 340 °C, while that the total inorganic carbon program has started from 530 °C until 1000 °C (1 min) and the peak temperature was 775 °C. One of the problems in the present study was how to calculate the total carbon content in the HA sodium salt. Probably the carbonyl groups were not free (in the form –COOH) but were rather in the form of –COONa.

The evolved CO_2 would react with $-COONa$ and CO_3^{-2} would be formed (Voskamp 2004) giving reason for the high value of pH neutral HA solution at (6.5 - 7). It has been found that the total organic carbon was 11.69 % and the total inorganic carbon was 27.82 %, but the real percentage of total organic carbon is (39.79 and 39.24%), such that f_{oc} is found to be 0.395 g/g.

Table 3.2 Elemental analysis of Aldrich HA.

Sorbant	%TOC	% TIC	%TC	%H	%N	%S	f_{oc} in g/g
Aldrich HA	11.70	28.09	39.79	5.24	0.37	0.38	0.3979
	11.69	27.55	39.24	5.68	0.42	0.64	0.3924

Preparation of the humic acid solution

The first step has been the preparation of a buffer solution at three different pH [25 °C± 2] values (4 -10± 0.2). The buffer solution with pH 4 has been prepared by the addition of 0.02 mM HCl to 10 mM $C_8H_5KO_4$ while the buffer of pH 10 has been prepared by the addition of 3.36 mM NaOH to 96.64 mM Na_2HPO_4. The buffer pH 7 has been prepared by the addition of 24.4 mM HCl to 75.6 mM Na_2HPO_4. 200 mgL^{-1} of NaN_3 as an antimicrobial agent (Ilani et al. 2005) was added to the buffer solution, and then pH values were measured by a pH-meter. The constant ionic strength of the HA solution was adjusted by 0.01-1 M NaCl solution as a regulator agent. The electrical conductivity of the buffer solution was measured by the conductivity-meter and was found 11.33±2 mScm^{-1}. Different concentrations of HA were prepared and diluted with various concentrations, and then the pH values were controlled again. The stock solutions were left in the dark in brown bottles. After this preparation, these HA solutions have been ready for use in sorption experiments.

Preparation of standard solutions of the organic chemicals

The prepared standard solutions of the organic solutes depended on their water solubility values and their 1-octanol-water partition coefficients. Furthermore, consideration has been given to ensure that the percentage of the solvent has not been more than 0.22% (V/V) to avoid the co-solvency effect (Wauchopa et al. 1983 and Yu et al. 2006). 2-Propanol, methanol and acetonitrile have been used depending on the chemical and physical nature of the organic substances and then those stock solutions have been kept at about 5-8 °C in the refrigerator.

3.2.3.1 Direct immersion mode SPME-GC-MSD (in full vial without headspace)

A set of first experiments have been carried out at the beginning of the study to determine the required HA content for sorption of different organic sorbates and also regarding the literature data of log K_{oc} and values of log K_{ow}. Two sets of calibrations of each analyte have been carried out. Both have been obtained by spiking different concentrations of dissolved analyte in 2-propanol and buffer reference solution. Both of calibration series have been injected to the GC system. The first calibration determined the absolute amount of analyte to estimate the fibre-caused depletion (negligible depletion-SPME should not be going beyond 5% according to Vaes et al. 1997). The second calibration estimated the freely dissolved amount of the analyte in the water phase. The used fibres have shown a negligible depletion for all investigated analytes. The direct immersion mode of full vial extraction in sorption measurements has been performed by three different types of fibres as the nd-PDMS 100 µm, nd-PA 85 µm, and nd-PDMS-DVB 65 µm. The sorption experiments have been carried out with a constant HA concentration-water volume ratio and a variation of HA-analyte concentration ratio for each analyte separately. Sorption experiments have been performed by batch equilibrium method (OECD106 2000, OECD121 2001 and von Oepen et al. 1991) in 10, 14 and 20 mL Gerstel vials which have been filled fully with HA solution spiked with some µL from standard solution of the selected solutes and covered by using screw-cap vials (minimal head space < 0.5 %) with septum. The samples of buffer reference and HA contained different amounts of investigated analytes in different buffer solution pH (4, 7, and 10±0.2), with NaN_3 as antimicrobial agent and NaCl as constant ion strength regulator. The blank vials with reference buffer contained the organic sorbates without HA and passed the same procedures (equilibrium, water-desorption, nd-SPME fibres, and GC analysis) as the sorption vials of HA. The equilibrium attained through continuous agitation of 24-48 h depending on the differences of volatility and hydrophobicity of the analytes. Some of sorption experiments have been done by continuous agitation inside the incubator at different equilibrium temperatures 4-45 ± 2 °C. For the full mode vial analysis, the SPME fibre has been exposed to the aqueous phase for a constant extraction time in different temperatures of agitator in the same speed of 250 rpm, depending on the extraction kinetic of the fibre and the chemical nature of analytes. The extraction time has been varied between 10 and 60 min. After extraction at each temperature the fibre has been injected to the GC-MSD splitless injector port for desorption time between 1 and 3 min at 250 °C. The samples have been analysed by variation of the oven temperature program for each selected analyte as illustrated in Appendix 9.11 and 9.12. The analysis has been repeated two or three times for each of the test vials containing the HA.

3.2.3.2 Headspace mode HS-SPME-GC-MSD (in half-filled vial)

The same extraction kinetics as mentioned above have been applied in headspace mode (Pawliszyn 2001, Krutz et al. 2003, ter Laak et al. 2005, Kolb et al. 1992 and Kolb et al. 2006) sorption experiments that have been performed using batch technique in 20 mL Gerstel vials. The vials have been half fully filled with HA solution spiked with some of the µL from standard solution of the selected solutes and covered by using Gerstel screw-cap vials with septum. The sorption experiments have been performed in pH7±0.2 with NaN_3 as antimicrobial agent and NaCl as constant ion strength regulator at different equilibrium temperatures (5, 25, and 45 ± 2 ^oC). The equilibrium has been attained through continuous agitation between 24 and 48 h for all the sorbates. For a half full mode vial analysis (with free headspace), the SPME fibre has been exposed to the gaseous phase for a constant extraction time at different temperatures of the agitator with the same speed of 250 rpm. The extraction time has been varied between 10 and 60 min. The samples have been analysed using GC-MSD with variation of the oven temperature program for each selected analyte. The analysis has been repeated two or three times for each of the test vials containing the HA.

3.2.4 GC-MSD analysis

Three different GC-MSD systems have been used in a chromatographic analysis. The first was an HP 5890 Series gas chromatograph with an HP 5971 series mass selective detector (Hewlett Packard) combined with an autosampler (Varian CP-8200). The stationary phase of capillary column was CP-Sil 8 CB-MS 50 m x 0.32 mm x 0.12 µm (5 % diphenyl + 95 % dimethylsiloxane copolymer unpolar column). Helium was the carrier gas with a constant pressure of 60 k Pa and 2 min splitless. The second was an Agilent Technol 6890N series gas chromatograph with an Agilent 5973N series mass selective detector combined with Gerstel multipurpose autosampler. The stationary phase of the capillary column was HP-5MS 30 m x 0.25 mm x 0.25 µm (5% phenyl + 95% dimethylsiloxane). Helium was the carrier gas with a constant flow of 1.1 mL min^{-1} and 2 min splitless. The third was Agilent Technol 7890 A series Gas Chromatograph with Agilent 5975C series mass selective detector combined with Gerstel multipurpose autosampler. The stationary phase of capillary column was HP-5MS 30 m x 0.25 mm x 0.25 µm film thickness (5% phenyl + 95% dimethylsiloxane). Helium was the carrier gas with a constant flow of 1.1 mL min^{-1} and 2 min splitless. Different temperature programs have been used for different substances according to their physical and chemical properties as listed in Appendix 9.12.

The analyte shave been detected and quantified by electron ionization (EI)-MS using NIST 98. The ionization of the gas current resulted in ions positively mode at ca. 2012 V, 183°C and 61 Pa. For the determination of the retention time of an analyte, the GC-MSD in the *SCAN* mode has been operated (reception of the total ion current, reception area, 50-550 ms/z) as shown in Figure 3.8. For the identification of analytes over the retention scale of the gas-chromatograms, the fragment ions and the molecule ion of each analyte were compared out of the total ion current to the database NIST 98 as shown in Figure 3.9. Additional the total ion current has been examined on the fragment ions itemised, the GC-MSD-courses to the concentration determination have been carried out in the single ion mode (single ion, *SIM* mode) as shown in Figure 3.10. Those at the same time for detecting ions for each analyte have been shown in Appendix 9.11.

For K_{oc} determination, the distribution balance should be examined for each solute between more aqueous and solid phase (HA). The concentration determination of the solute in aqueous phase has taken place using accompanied SPME technology with GC-MSD. First set of tests have been carried out in full direct mode SPME at room temperature with all analytes in order to determine the detectable concentration area (in the *SCAN* mode) and/or the order of magnitude for a later calibrate straight line and the retention times as well as identities of the molecules. The concentration of freely dissolved amount of analyte in water phase measured in *SIM* mode MSD, whereas the difference in mass intensities have displayed the sorption effect by HAs as shown in Figure 3.10.

The limit of detection (LOD) and limit of quantitation (LOQ) have been considered in the selection of test values as shown in Figure 3.11. Out of the values of all reference samples, a calibrated straight line has been generated and statistically tested. Using this, values which were outside of the confidence interval, have been left out. With the help of this statistical test an allowable calibrate straight line can be used for further calculation. The limit of quantitation served as a lower end criterion for the use of the concentration values of sorption experiments to the elevation of the K_{oc}. Subsequently the average of all K_{oc} allowable after this criterion has been determined and their logarithmic forms (log K_{oc}).

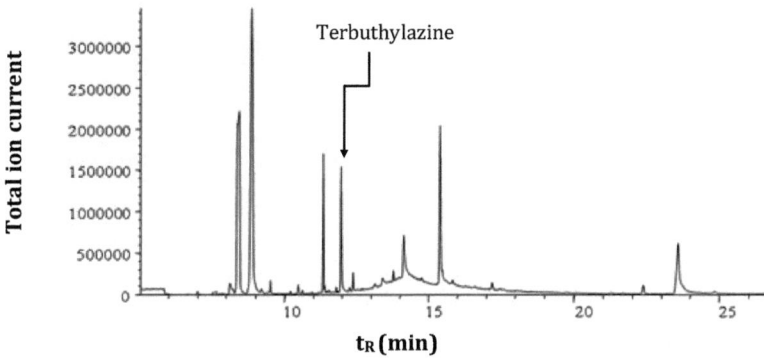

Figure 3.8 Chromatogram of terbuthylazine at t_R 11.33 min after 15 min SPME (85 µm Polyacrylat) via headspace mode in room temperature (25 ± 2 °C) for aqueous solution (181.1 µgL^{-1} in buffer solution of pH = 7± 0.2) with *SCAN* mode GC-MSD.

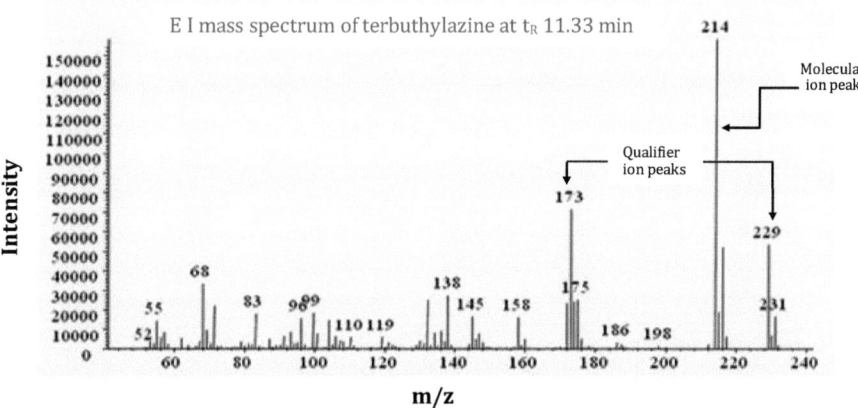

Figure 3.9 Chromatogram of electron ionization (EI)-mass spectra of terbuthylazine at t_R 11.33 min with *SCAN* mode GC-MSD, m/z were (214, 173, 229) for the molecular target ion and qualifier ion peaks (NIST 98).

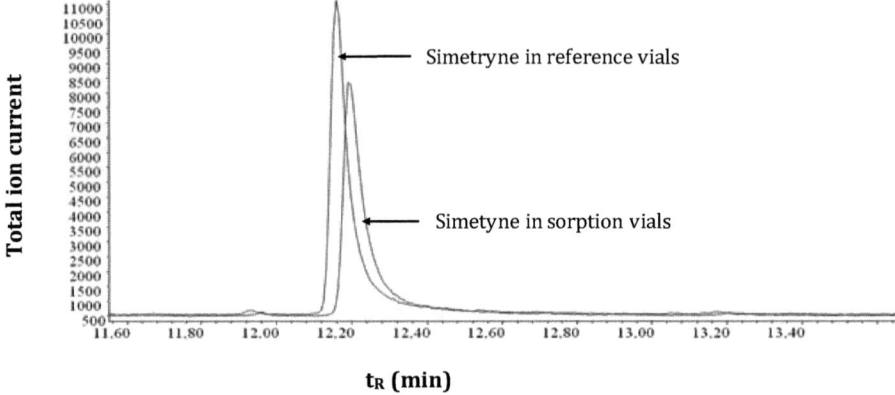

Figure 3.10 Chromatogram of simetryne at t_R 12.14 min in a reference as well as in sorption experiment with 600 mgL^{-1} HA after 15 min SPME (85 µm PA) via HS-SPME in room temperature (25 ± 2 °C) for aqueous solution (144.9 µgL^{-1} in buffer solution of pH7± 0.2) with *SIM* mode GC-MSD.

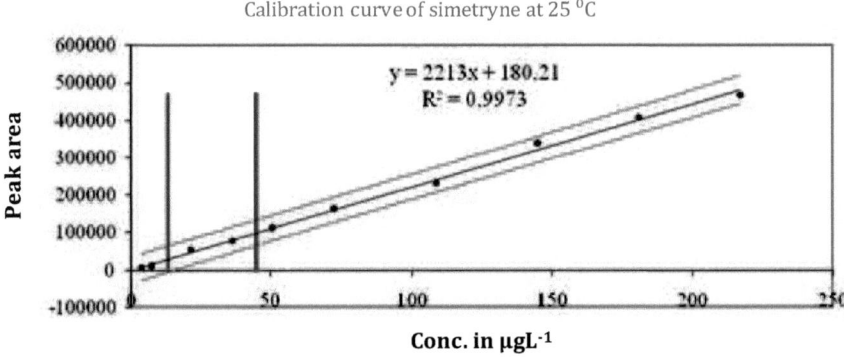

Figure 3.11 Graphical plot of data for simetryne test series in 25 ± 2°C. ——— Black line is linear line. ——— Green line is limit of detection (13.5 µgL^{-1}) and ——— blue line is limit of quantitation (44.9 µgL^{-1}) both calculated from the calibration curve according to DIN 32645, (2011-8). ——— Red line is confidence interval 95% (Burke 2007).

3.2.5 Determination of K_{oc}

It is well known that in the aqueous system, a linear relationship exists - at equilibrium - between the concentration of a contaminant in solid phase and in aqueous phase. In other words, when the concentration of a contaminant in water is increased, the concentration of that contaminant in the soil/sediment will increase by a constant factor. This linear relationship is expressed mathematically via the solute distribution coefficient which is the ratio of the solute concentration in soil C_s to the solute concentration in water C_w.

$$K_d = \frac{C_s}{C_w} \qquad (3.2)$$

Where: K_d is the solute distribution coefficient, C_s is the solute concentration in solid phase and C_w is the solute concentration in aqueous phase (Georgi 1998) Knowing the concentration of a contaminant in one compartment (either water or HSs) thus helps in predicting the concentration of a contaminant in the other compartment. Since the contaminant sorption occurs predominantly by partitioning into the HA, it is more useful to express the distribution coefficient in terms of the total organic carbon (f_{oc}).

$$C_w = \frac{moles}{kg_{solids}} \qquad (3.3)$$

The units on K_d are defined as, $K_d = \left[\frac{\frac{moles}{kg_{solids}}}{\frac{moles}{L_{water}}} \right] = \frac{L_{water}}{kg_{solids}} \qquad (3.4)$

In equation (3.5) K_{oc} is the partition coefficient normalized to the organic carbon of the HA and f_{oc} is the total organic carbon fraction of the HA. In this study K_{oc} is considered to be the HA-water partition coefficient (normalized to the organic carbon).

$$K_{oc} = \frac{K_d}{f_{oc}} = \frac{C_s}{C_w f_{oc}} = \frac{C_{Sorbiert}}{C_{Free}} \cdot \frac{1}{f_{oc}} \qquad (3.5)$$

K_{oc} can be determined by using two methods: The first is from the calibration curve of the recorded peak area for the concentration of the analyte in the aqueous phase without humic matrix, and the second is by direct calculation from the signal response (value of peak area) of the SPME-GCMSD.

Method I for K_{oc} determination

It was supposed that, the fibre takes up all the mass (which can be transferred from the vial to the fibre then injected to the GC-MSD) due to the negligible depletion percentage of the fibre and those low K_{ow}. The substance which has a high K_{ow}, can easily be taken up by the fibre. In this case, a high precision peak should be expected, and the substance which has a low K_{ow}, needs a high matrix concentration to obtain a high precision peak because of the difficulty of transfer to the fibre. In this method, the calibration curve depends on the water solubility of the analyte, prepared concentration range and the recorded peak areas. The concentration of the analytes was compared by using the calibration curve with and without humic matrix after equilibrium time had been reached. The residual free concentration of the analyte in aqueous phase of HA-water system was then calculated. As shown in equation (3.6), K_{oc} can be determined by adding some factors related to sorption and the sampling process according to the literature (Poerschmann et al. 1997, Georgi 1998 and Mackenzie et al. 2002). The derivations of equations (3.6) and (3.9) have shown in Appendix 9.19.

$$K_{oc} = \frac{C_{ref} - C_{matrix}}{C_{matrix} \cdot C_{humic} \cdot f_{oc}} \cdot \left[\frac{K_{fibre} V_{fibre} + K_H V_{gas}}{V_w} + 1 \right] \quad (3.6)$$

C_{ref}: Initial concentration of the solute $\left[\frac{mg.10^{-3}}{M.wt.L} = mole/L\right]$,

C_{matrix}: Free concentration of the solute after sorption /or the concentration of solute in aqueous phase $\left[\frac{mg.10^{-3}}{M.wt.L} = mole/L\right]$,

$C_{ref} - C_{matrix}$: Bonded sorbate concentration in solid phase $\left[\frac{mg.10^{-3}}{M.wt.L} = mole/L\right]$,

C_{humic}: HA concentration (mg.10⁻⁶/L= kg/L),

f_{oc}: Total organic carbon fraction of the HA (kg/kg),

K_{fibre}: Fibre (stationary phase) water partition coefficient $K_{PDMS-water}$, (Sprunger et al. 2007).

V_{fibre}: Fibre volume (L),

K_H: Henry's law constants (dimensionless gas/aqueous) at 25 °C (Kühne et al. 1997 and Kühne et al. 2005).

V_{gas}: Headspace volume ≅ air volume (L),

V_m: Total aqueous volume (L).

In the case of SPME direct immersion-mode full vial analysis (without head space, the term $K_H V_{gas}$ can be neglected, therefore $K_H V_{gas}/V_w$ is a negligible term (note the very diminished factor)

$$K_{oc} = \frac{C_{ref} - C_{matrix}}{C_{matrix} \cdot C_{humic} \cdot f_{oc}} \cdot [K_{fibre} V_{fibre} + 1] \qquad (3.7)$$

If the SPME fibre volume partitioning into the fibre is sufficiently small, the term $K_{fibre}V_{fibre}$ can be neglected (note that V_{fibre} [L⁻⁶] $<<$ V_w [L⁻³])

$$K_{oc} = \frac{C_{ref} - C_{matrix}}{C_{matrix} \cdot C_{humic} \cdot f_{oc}} \qquad (3.8)$$

Method II for K_{oc} determination

Direct determination of K_{oc} can be carried out by applying equation (3.9) (Georgi et al. 2002) and using the signal response of the GC-MSD value of peak area of the analyte concentration. This method was helpful with the samples which had very low water solubility and very high 1-octanol-water portioning coefficient.

$$\frac{SPME_{(ref)Signal}}{SPME_{(matrix)Signal}} = \frac{a_{(ref)}}{a_{(matrix)}} = 1 + K_{oc} f_{oc} C_{humic} \qquad (3.9)$$

$$K_{oc} = \left(\frac{C_{ref}}{C_{matrix}} - 1 \right) \cdot \frac{1}{f_{oc} C_{humic}} = \frac{C_{ref} - C_{matrix}}{C_{matrix}} \cdot \frac{1}{C_{humic} f_{oc}} \cdot \left[\frac{K_{fibre} V_{fibre} + K_H V_{gas}}{V_w} + 1 \right] \qquad (3.10)$$

The concentration can be replaced by peak area from SPME-GCMSD signal for both reference and matrix vials as in equation (3.11). A_{ref} is the peak area of solute without HA in blank vial while A_{matrix} is that peak area of free amount after addition of HA.

$$K_{oc} = \left(\frac{A_{ref}}{A_{matrix}} - 1 \cdot \frac{1}{f_{oc} C_{humic}} \right) = \frac{A_{ref} - A_{matrix}}{A_{matrix}} \cdot \frac{1}{C_{humic} f_{oc}} \cdot \left[\frac{K_{fibre} V_{fibre} + K_H V_{gas}}{V_w} + 1 \right] \quad (3.11)$$

In the case of SPME Direct mode full vial analysis (without head space) equation (3.11) will be replaced by equation (3.12) after neglecting the term $K_H V_{gas}/V_w$

$$K_{oc} = \left(\frac{A_{ref}}{A_{matrix}} - 1 \cdot \frac{1}{f_{oc} C_{humic}} \right) = \frac{A_{ref} - A_{matrix}}{A_{matrix}} \cdot \frac{1}{C_{humic} f_{oc}} \cdot \left[\frac{K_{fibre} V_{fibre}}{C_{humic} f_{oc}} + 1 \right] \quad (3.12)$$

If SPME fibre volume partitioning into the fibre is sufficiently small, the term $K_{fibre}V_{fibre}$ can be neglected (note that V_{fibre} [L⁻⁶] $<< V_w$ [L⁻³]) and so equation (3.13) will be obtained as follows:

$$K_{oc} = \left(\frac{A_{ref}}{A_{matrix}} - 1 \cdot \frac{1}{f_{oc} C_{humic}} \right) = \frac{A_{ref} - A_{matrix}}{A_{matrix} C_{humic} f_{oc}} \quad (3.13)$$

The log for all values of K_{oc} has been calculated from the results of both of SPME-GSMS analysis.

3.3 Comparison with mean log K_{oc} in literature

The experimental log K_{oc} values in current study measured at 25±2 °C have been compared with three types of mean values of log K_{oc} in selected literature (1995-2011) data, related to different sorbents, firstly related to different HAs as OM matrix, secondly related to various soils and sediments, and thirdly to the non-specific data which were collected from references without indication of their original sorbent types (different OM). Those data have been compared with the experimental values in the current study. Log K_{oc} of forty-seven substance have been found in literature while those of fourteen substances have not been measured before as shown in Appendix 9.16 (H4, G2, E2, E4, K3, D1, H1, H3, K4, L1, L2, L4, NI, and O2).

3.4 QSAR modelling validation and predictability

Prediction modelling of log K_{oc} values for various categories of chemical classes in different experimental conditions such as variation of pH and temperature has been carried out by two types of QSAR modelling. A set of statistical significance tests have been carried out. Predicted models have been validated by advanced calculated statistical parameters. It will be discussed in details in the next paragraphs.

3.4.1 Log K_{ow}-log K_{oc} correlationship

The classical approach prediction of sorption coefficients normalized to organic carbon log K_{oc} based on the solute hydrophobicity (in terms of 1-octanol-water partition coefficient) has been used 1-octanol as a surrogate for organic carbon in organic suspended matter and soils. It has been developed accordingly to the following equations: (Karickhoff 1981, Sabijić 1987, Sabijić et al. 1995, Galwik et al. 1997)

$$K_{oc} = 0.41\, K_{ow} \qquad (3.14)$$

and its logarithm form $\qquad \log K_{oc} = \log K_{ow} - 0.39 \qquad (3.15)$

Prediction of log K_{oc} based on K_{ow} can only be applied to neutral hydrophobic compounds whereas the dominant process was absorption into HA as OM sorbent. The classical model could not provide good predictions for ionizable chemicals that can undergo to dissociation or protonation associated with specific intermolecular interactions.

The experimental log K_{ow} values of the probe compounds have been obtained using KOWWIN™- Episuite™ v.4 software (Meylan 2004). The calculated log K_{oc} values from KOCWIN™-Episuite™ software (Meylan et al. 1992 and 2002) were estimated by using two different calculation methods, the first based on log K_{ow} common model (Karickhoff 1981 and Doucette 2003) and the other based on the increment indices-fragmentation method from molecular structure (Tao et al. 1999 and Schüürmann et al. 2006). The linear regression fit of log K_{ow}-log K_{oc} of the investigated probe has not only been obtained for different chemical classes but also in different sorption conditions.

3.4.2 Linear Solvation Energy Relationship (LSER)

The LSER equation of Abraham (Abraham et al. 1987, 1991, 1993, 1994 a, b, and 2004) has been used to model the solvation-dependent properties of organic solute between two bulk phases. Recently, LSER has been widely used to predict the equilibrium partition coefficients in sorption processes; therefore the modified equation form includes both phase and solute parameters related to particular types of intermolecular interactions.

$$\log K_{oc} = eE + sS + aA + bB + vV + c \tag{3.16}$$

E, S, A, B, V, are solute descriptors while e, s, a, b, v, are the counterpart phase (solvent) parameters and c is regression constant.

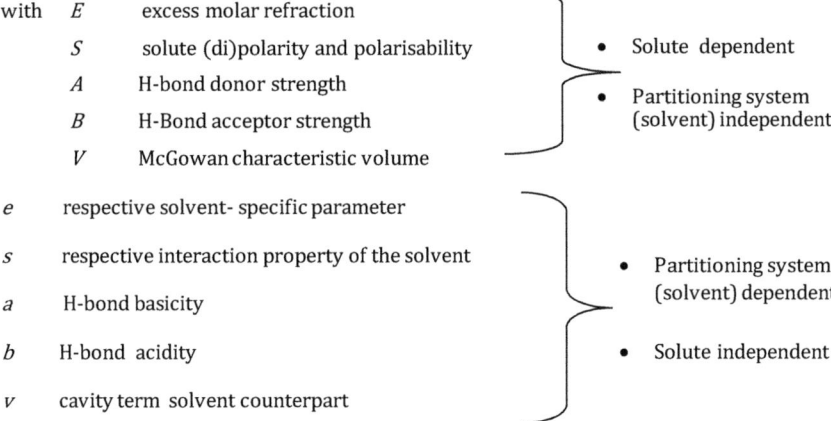

with
- E excess molar refraction
- S solute (di)polarity and polarisability
- A H-bond donor strength
- B H-Bond acceptor strength
- V McGowan characteristic volume

- e respective solvent- specific parameter
- s respective interaction property of the solvent
- a H-bond basicity
- b H-bond acidity
- v cavity term solvent counterpart

- Solute dependent
- Partitioning system (solvent) independent

- Partitioning system (solvent) dependent
- Solute independent

In our modelling, two types of solute descriptors have been used. Firstly, S, A, and B_0 (wet system which saturated with strong H-bond donor as humic-water system) have been estimated experimentally (Poole et al. 1999 and Nguyen et al. 2005) by different spectroscopic methods such as GC-MSD (Abraham et al. 1999), RP-HPLC (Zissimos et al. 2002 and Vitha et al. 2006), H^1–NMR (Basso et al. 2007), and FTIR (Kamlet et al. 1976), secondly they have been estimated by the increment method (Platts et al. 1999). The McGowan volume could be calculated from the structure. The excess molar refraction E, has been calculated regarding to the refractive index η and McGowan volume (Abraham et al. 1987).

$$\text{MR}_X = 10\left[\frac{(\eta^2+1)}{(\eta^2+1)}\right].V \qquad (3.17)$$

η can be estimated by ACD software and it has been measured normally at 20 °C.

$$E = \text{MR}_X - (\text{MR}_X)_{\text{Alkan}} \qquad (3.18)$$

$$E = \text{MR}_X - 2.83195\,V + 0.52553 \qquad (3.19)$$

McGowan volume *V* can be estimated by fragmentation method according to the number of the elements atoms and the bonds for each probe compound as displayed in Table 3.3.

Table 3.3 An example of atomic volume *V* for some different elements in cm^3 mol^{-1}.

Atom	V	Atom	V
H	8.71	F	10.48
C	16.35	Cl	20.95
O	12.43	Br	26.21
N	14.39	I	43.53

$$V = \sum(ni.Vi) - m.6.56 \text{ cm}^3 \text{ mol}^{-1} \qquad (3.20)$$

The physical properties of all solutes were collected from different databases and the missing data were summed up and calculated by using "UFZ in-house version of ChemProp" (Schüürmann et al. 1997). The LSER-Abraham experimental descriptors (Abraham et al. 1987, 1994, and 2004) of solutes and the calculated descriptors (Platts et al. 1999) were collected from ChemProp software (Schüürmann et al. 1997 and 2007). Prediction methods of log K_{oc} values of probe set have been carried out by two types of QSAR approaches. Firstly the common approach of log K_o-log K_{ow} correlation (Karickhoff 1981, Sabijić et al. 1995, Galwik et al. 1997 and Doucette 2003). The experimental log K_{ow} values have been obtained using KOWWIN™ - Episuite™ v.4 software (Meylan et al. 1992). The linear regression fit of log K_{ow}-log K_{oc} of the investigated probe has been obtained for different chemical classes. Henry's law constant at 25 ±2 °C has been collected from Environmental Science-Interactive physprop database of Syracuse Research Corporation (SRC) and the missing data has been calculated using ChemProp software and converted to dimensionless values. The log K_{oc} values at 25 °C from KOCWIN™- Episuite™ v.4 (Meylan et al. 1992 and 2000) were calculated by two different calculation methods. The first method based on log K_{ow} common model while the second based on the molecular topology. The second approach of log K_{oc} prediction using LSER with multilinear regression analysis. Log K_{oc} data were calculated from the phase parameters *e, s, a, b, v* and *c*. These parameters have been calculated from the regression analyses of both Abraham and Platts LSER solute descriptors at 25 ±2 °C using STATISTICA software at confidence limit 95%.

3.4.3 Comparison of experimental and calculated log K_{oc} from different prediction methods

The experimental log K_{oc} data in the current study measured at 25±2 °C and pH at 7±0.2 has been compared with the calculated log K_{oc} using different prediction methods of the in literature as shown in Appendix 9.17. The following methods have been included in the comparative analyses:

i) A two-dimensional molecular structure model by Schüürmann et al. (2006) is a multilinear model containing the variables molecular weight, bond connectivity, molecular E-state, an indicator for non-polar and weakly polar compounds, and 24 fragment corrections representing polar groups.

ii) Log K_{oc} data have been calculated by KOCWIN software using first-order molecular connectivity indices and 27 polar fragment corrections also using the experimental database of log K_{ow}.

iii) A molecular connectivity indices model by Sabljic' et al. (1995) is employing 19 different regression equations based on log K_{ow} and one $^1\chi$ based regression to calculate log K_{ow} from molecular structure using KOWWIN.

iv) A fragment constant model by Tao et al. (1999) is consisting of 74 fragment values and 24 structural correction factors.

v) LSER model by Nguyen et al. (2005) is employing water-calibrated solute basicity values and LSER model by Poole et al. (1999) is employing octanol calibrated solute basicity values. Some additional comparisons were done with the recent published models such as Franco and Trapp model (2008)-weak acid model after consideration $pK_a \simeq 40$ or 50, it was only applied for unionized fractions as shown in Appendix 9.18.

vi) The new models of LSER approach were applied in the current study for both solute experimental and calculated descriptors such as Endo et al. (2009a), Bronner et al. (2010a), and Neale et al. (2012). For the validation of all methods, r^2 and q^2 were calculated, where r^2 quantifies the calibration performance and q^2 quantifies the prediction performance.

3.4.4 Statistics

The regression of prediction methods of log K_{oc} have been calculated and calibrated using different statistics methods using different software programs such as STATISTICA, "UFZ in-house version of ChemProp", and R.

3.4.4.1 Regression analysis

a) Multilinear regression

With multilinear regression, we assume that the log K_{oc} dependent values, y_i, depends linearly on several independent variables x_i, i.e., *e E, s S, a A, b B₀* and *v V*. However, for the estimation of phase parameters *e, s, a, b, v* and *c*, the multilinear regression coefficients of solute descriptors, those have been calculated using the mulitlinear regression function of STATISTICA software at confidence limit 95%. Phase parameters have been estimated for both Abraham and Platts LSER solute descriptors from exp. log K_{oc} data at different pH 4, 7, and 10 as well as at 5, 25, and 45 $\pm$ 2°C. Regression coefficients of different models have been estimated and calibrated using significance test at confidence limit of 95%. Significance tests have been carried out using both T-test for normally distributed populations of the values between two different groups and F-values calculated using Fischer-test for multilinear regression at confidence interval 95%.

b) Leave-one-out cross-validation

The cross-validation has been used to estimate the power level of fit of a model to a data set that is independent of the data that been used to train the model. Leave-1-out cross-validation (Schüürmann et al. 2008) has been calculated for *N* subset containing *N-1* compounds which would be the new subset to calibrate the prediction of property of each compound. The mathematical formula of predictive regression with leave-1-out-cross-validation has been illustrated in Appendix 9.21. The performance of QSAR new proposed models has been calibrated by leave-one-out cross-validation (LOOCV) method. Moreover, it has been used to compare between the performances of different methods of log K_{oc} prediction such as Franco and Trapp established model (Franco et al. 2008 and 2009) and a new proposed predicted models in the current study as shown in chapter 4.2.

3.4.4.2 Performance parameters

Calibration and prediction performance of the proposed models from regressions have been evaluated using the statistical parameters, their mathematical formula have been displayed in Appendix 9.21. The prediction performance parameters have been calculated at 95% confidence limit to calibrate the new proposed models and to compare with established models such as r^2 squared correlation coefficient (Moore 2004 and Kessler 2007), q_{cv}^2 (predictive squared correlation coefficients using leave-one-out cross-validation as described in Schüürmann et al. (2008), *rms* (root-mean-square error of correlation), rms_{cv} (root-mean-square error of prediction using leave-one-out cross-validation), *bias* (systematic error), *me* (mean error), *mne* (maximum negative error), *mpe* (maximum positive error) and *F*- values have been calculated at α = 0.05 (Doerffel 1966).

4. Results and Discussion

4.1 A closer look at sorption of xenobiotics to humic acid in neutral water

4.1.1 Introduction

The aim of this part of the study was to establish a prediction model for sorption of non-polar and polar hydrophobic organic compounds from water to dissolved HA. The proposed model has been calibrated with data for sixty-one compounds (polar and non-polar) from different chemical classes covering phenols, heterocyclic compounds, organochlorines and PAHs. There is a lack of literature providing data on the prediction performance of each existing method in the prediction of log K_{oc} for different chemical domains separately. Therefore, one of the aims of this study was to compare between different prediction methods depending on their applicability domain and to test the applicability of the LSER approach on a diversity of calibration dataset that includes neutral compounds, acids and bases as well as the possible applicability for different chemical classes. In addition, it was interesting in the reliability of a combination of K_{ow} and LSER approaches to be more effective predictors of K_{oc} for those types of chemicals. In this study, the batch technique has been utilized to investigate the nature of the sorption process onto HA and investigate the intermolecular interaction parameters which are dominant in the sorption process as well as the nature of binding. Experimental log K_{oc} values of the dataset were determined at pH 7 ± 0.20 and 25 ± 2 °C. Different types of organic solutes were chosen to be investigated with a large range of hydrophobicity in terms of log K_{ow} from 0.65 to 8.27 including PAHs, organochlorines, phenols and heterocyclic compounds by using negligible depletion (nd)-SPMS-GC-MSD as an analytical tool.

4.1.2 Comparison with mean log K_{oc} in literature

Log K_{oc} data (f_{oc} of used HA is 0.395 g/g) of substances have been measured in headspace and direct mode as shown in Appendix 9.10. Log K_{oc} values were determined from equations (3.6) and (3.8) as described in chapter 3. Table 4.1.1 shows the exp. log K_{oc} data and log K_d for all sorbents, and Appendix 9.16 shows the comparison of the experimental log K_{oc} in current study with the mean value of the experimental log K_{oc} in literature. Table 4.1.2 shows the results of the linear regression of the comparison with literature data. Regression coefficients are slightly different for HAs, soils and sediments with a high standard deviation. The determined log K_{oc} values related to different carbon contents of various organic matter in addition to different compositions, carbonyl, phenolic, amino group content in each sorbent. Whereas the wide range of log K_{oc} has been expected due to different solute-sorbent (solvent) interactions including different contribution of intermolecular interactions, the difference will be associated with changes in solvent parameters as described before in the LSER approach. Fourteen substances have not been measured in the literature as illustrated in Appendix 9.16, compounds listed in scheme 3.1 (H4, G2, E2, E4, K3, D1, H1, H3, K4, L1, L2, L4, N1, and O2).

Figure 4.1.1 shows that the not-specified data are considered as collected data reflecting various types of sorbents as mixture of HAs, soils and sediments data. The sorption of HAs in literature is represented by [0.96±0.12 log $K_{oc\,(exp.)}$ + 0.08±0.45] while the sorption of soils and sediments is represented by [0.92±0.12 log $K_{oc\,(exp.)}$ + 0.25±0.46] This very small difference in sorption behaviour may be attributed to similar types of molecular interactions. From a statistical point of view, all intercepts are insignificant but show a small increase in soil sorption capacity.
Soil sorption normalized to organic carbon capacity relevant to soils and sediments is higher than that of HAs by 0.17 log units (difference between both intercepts) due to stronger molecular interaction in soils and sediments than in HAs. A slight difference of sorption capacity may be attributed to the difference in organic carbon contents and characteristic differences of the hydrophobic soil phase and the influence of mineral surfaces (Chiou et al. 1983). It indicates that the sorption can be expressed as a function of the organic carbon content of the sorbent (Doucett 2003) without neglecting the differences in organic matter properties and the magnitude of non-organic matter contributions to the sorption process.

Table 4.1.1 The experimental values of sorption coefficients for different organic substances at 25 ± 2 °C and pH 7 ± 0.2 (f_{oc} of used Aldrich HA is 0.395 g/g).

No.	Substances	CAS No.	Log K_{ow}	pK_a	$K_{d(mean)}$ [L/kg]	Log K_d ± Err	$K_{oc(mean)}$ [L/kg$_{oc}$]	Log K_{oc} ±Err
				Anilines				
1	Aniline	62-53-3	0.90	4.60	140	2.15±0.39	355	2.55±0.39
2	4-Chloroaniline	106-47-8	1.83	3.98	1613	3.21±0.06	4074	3.61±0.06
3	2,3,5,6-Tetrachloraniline	3481-20-7	4.10	-1.40**	2557	3.41±0.05	6456	3.81±0.05
			R-benzene (R= H, CH$_3$, 3(CH$_3$), and CHO)					
4	Benzene	71-43-2	2.13	n.a.	21	1.33±0.03	54	1.73±0.03
5	Benzaldahyde	100-52-7	1.48	14.90	217	2.34±0.24	548	2.74±0.03
6	Toluene	108-88-3	2.73	n.a.	107	2.03±0.82	270	2.43±0.82
8	1,2,4-Trimethylbenzene	95-63-6	3.63	n.a.	945	2.98±0.16	2387	3.38±0.16
			Halobenzenes					
7	Fluorobenzene	462-06-6	2.27	n.a.	81	1.91±0.13	205	2.31±0.13
9	1,3,5-Trichlorobenzene	108-70-3	4.19	n.a.	415	2.62±0.11	1047	3.02±0.11
10	1,2,3,4-Tetrachlorobenzene	634-66-2	4.60	n.a.	12494	4.10±0.03	31550	4.50±0.03
11	Pentachlorobenzene	608-93-5	5.17	n.a.	8314	3.92±0.02	20994	4.32±0.02
12	Hexachlorobenzene	118-74-1	5.73	n.a.	11161	4.05±0.07	28184	4.45±0.07
			Organochlorines & Biphenyls					
13	2-Fluorobiphenyl	321-60-8	3.96#	n.a.	2549	3.41±0.08	6438	3.81±0.08
14	PCB 180	35065-29-3	8.27#	n.a.	702399	5.85±0.05	1773734	6.52±0.05
15	PCB 202	2136-99-4	7.73	n.a.	567246	5.75±0.05	1432439	6.16±0.05
16	Aldrin	309-00-2	6.50	n.a.	67250	4.83±0.15	169824	5.23±0.15
17	Lindane	58-89-9	3.72	n.a.	2442	3.39±0.09	6166	3.79±0.09
18	2,4`-DDD	53-19-0	5.87#	n.a.	147763	5.17±0.02	373139	5.57±0.02
19	2,4`-DDE	3424-82-6	6.00#	n.a.	101821	5.01±0.04	257123	5.41±0.04

Table 4.1.1 Continued.

No.	Substances	CAS No.	Log K_{ow}	pK_a	$K_{d(mean)}$ [L/kg]	Log K_d ± Err	$K_{oc(mean)}$ [L/kg$_{oc}$]	Log K_{oc} ±Err
20	2,4`-DDT	789-02-6	6.79#	n.a.	52880	4.72±0.05	133536	5.13±0.05
21	4,4`-DDD	72-54-8	6.02	n.a.	41467	4.62±0.05	104715	5.02±0.05
22	4,4`-DDE	72-55-9	6.51	n.a.	41469	4.62±0.07	104720	5.02±0.07
23	4,4`-DDT	50-29-3	6.91	n.a.	111608	5.05±0.11	281838	5.45±0.11
			PAHs & Substituted PAHs					
24	Naphthalene	91-20-3	3.30	n.a.	672	2.83±0.21	1698	3.23±0.21
25	1-Nitronaphthalene	86-57-7	3.19	n.a.	312	2.49±0.04	10001	3.00±0.04
26	2,3-Dimethylnaphthalene	581-40-8	4.40	n.a.	1895	3.28±0.18	4786	3.68±0.18
27	Anthracene	120-12-7	4.45	n.a.	1769	3.25±0.05	4467	3.65±0.05
28	2-Aminoanthracene	613-13-8	3.43	4.32**	22269	4.35±0.05	56234	4.75±0.05
29	2-Methylanthracene	613-12-7	5.00	n.a.	12632	4.10±0.02	31900	4.50±0.02
30	Fluorene	86-73-7	4.18	n.a.	761	2.88±0.13	1922	3.28±0.13
31	2-Nitrofluorene	607-57-8	3.37	n.a.	1039	3.02±0.03	2624	3.42±0.03
32	1-Aminopyrene	1606-67-3	4.31	4.32**	26163	4.42±0.19	66069	4.82±0.19
33	Fluoranthene	206-44-0	5.16	n.a.	10907	4.04±0.06	27542	4.44±0.06
34	Benzo(a)pyrene	50-32-8	6.13	n.a.	106585	5.03±0.07	269153	5.43±0.07
35	6-Aminochrysene	2642-98-0	4.99	4.32**	86635	4.94±0.02	218776	5.34±0.02
			Phenols					
36	Phenol	108-95-2	1.46	9.99	122	2.09±0.20	309	2.49±.20
37	4-Chlorophenol	106-48-9	2.39	9.41	140	2.15±0.14	917	2.96±0.14
38	4-Nitrophenol	100-02-7	1.91	7.15	165	2.22±0.13	417	2.62±0.13
39	2,4-Dichlorophenol	120-83-2	3.06	7.89	154	2.19±0.03	390	2.60±0.03

Table 4.1.1 Continued.

No.	Substances	CAS No.	Log K_{ow}	pK_a	$K_{d(mean)}$ [L/kg]	Log K_d ± Err	$K_{oc(mean)}$ [L/kg$_{oc}$]	Log K_{oc} ±Err
40	2,4,6-Trichlorophenol	88-06-2	3.69	6.23	93	1.97±0.34	234	2.37±0.34
41	2,3,4,6-Tetrachlorophenol	58-90-2	4.45	5.22	314	2.50±0.30	794	2.90±0.30
42	Pentachlorophenol	87-86-5	5.12	4.70	218	2.34±0.43	549	2.74±0.43
43	2,4-Di-t-butylphenol	96-76-4	5.19	11.7	3870	3.59±0.12	9772	3.99±0.12
44	3-Phenylphenol	580-51-8	3.23	9.64	847	2.93±0.10	2138	3.33±0.10
45	4,4`-Isopropylidenediphenol	80-05-7	3.32	10.29** 10.93**	1091	3.04±0.22	2754	3.44±0.22
46	4-n-Hexylphenol	2446-69-7	4.52#	10.18**	3786	3.58±0.13	9560	3.98±0.13
47	4-n-Nonylphenol	104-40-5	5.76	10.15**	29356	4.47±0.05	74131	4.87±0.05
48	4-n-Octylphenol	1806-26-4	5.50#	10.15**	3750	3.57±0.13	9469	3.98±0.13
49	Triclosan	3380-34-5	4.76	7.80**	20309	4.31±0.17	51286	4.71±0.17
		Heterocyclic compounds						
50	Acridine	260-94-6	3.40	5.45	1701	3.23±0.06	4297	3.63±0.06
51	Pyridine	110-86-1	0.65	5.23	9	0.95±0.07	23	1.35±0.07
52	2-Methylpyridine	109-06-8	1.11	6.00	300	2.48±0.13	758	2.88±0.13
53	6-Nitroquinoline	613-50-3	1.84	3.24**	644	2.81±0.06	1771	3.25±0.06
54	Ametryn	834-12-8	2.98	4.10	198	2.30±0.23	501	2.70±0.23
55	Atrazine	1912-24-9	2.61	1.70	250	2.40±0.11	631	2.80±0.11
56	Prometryn	7287-19-6	3.51	4.05	307	2.49±0.05	776	2.89±0.05
57	Propazine	139-40-2	2.93	1.70	424	2.63±0.03	1071	3.03±0.03
58	Sebuthylazine	7286-69-3	2.61*	2.50**	307	2.49±0.28	776	2.89±0.28
59	Simetryn	1014-70-6	2.80	4.00	346	2.54±0.14	873	2.94±0.14
60	Terbuthylazine	5915-41-3	3.21	2.00	498	2.70±0.26	1257	3.10±0.26
61	Terbutryn	886-50-0	3.74	4.03**	287	2.46±0.03	724	2.86±0.03

Err is maximum logarithmic error of K_{oc} and K_d, it was calculated as shown in Appendix 9.20. (#) is estimated log K_{ow} by Epi suite ™ v.4 (KOWWIN software). (**) is estimated pK_a by ACD/pK_a v.12 software. (*) is experimental log K_{ow} from Environmental science-interactive physprop database of Syracuse Research Corporation (SRC). (n.a.) is not available. The exp. log K_{oc} values of underlined substances was calculated from the peak area (Method II in chapter 3) due to the very small peaks of their calibration curves. f_{oc} of used Aldrich HA is 0.3951 g/g (see chapter 3).

Chapter 4 - Results and Discussion

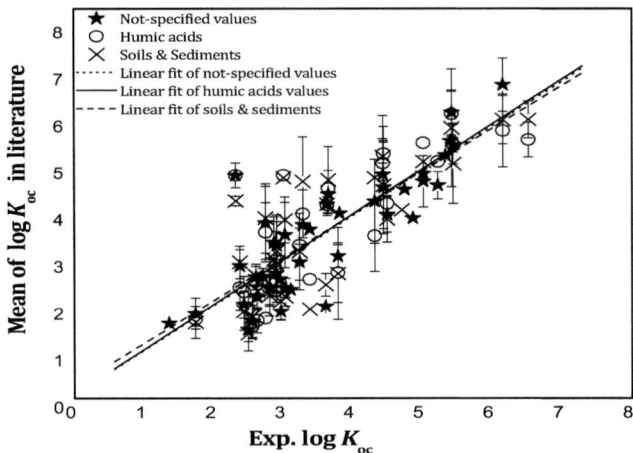

Figure 4.1.1 Comparison between mean values of exp. log K_{oc} in literature and exp. log K_{oc} in current study. Not-specified values are not relevant to specific sorbent.

Table 4.1.2 Results of linear regression of mean exp. log K_{oc} in literature versus exp. log K_{oc} in the current study.

Sorbent type	Regression equation	N	r^2	Sd
Humic acids	log $K_{oc\ (literature)}$ = (0.96±0.12) log $K_{oc\ (exp.)}$ + *(0.08 ±0.45)*	33	0.80	0.83
Not-specified	log $K_{oc\ (literature)}$ = (0.95±0.09) log $K_{oc\ (exp.)}$ + *(0.07±0.36)*	42	0.85	0.70
Soils & Sediments	log $K_{oc\ (literature)}$ = (0.92±0.12) log $K_{oc\ (exp.)}$ + *(0.25±0.46)*	35	0.79	0.83

a is slope and *b* is intercept. N is no. of points, sd is standard deviation, r^2 is squared correlation coefficient. 14 substances have not been measured in the literature as illustrated in Appendix 9.16, compounds listed in scheme 3.1 (H4, G2, E2, E4, K3, D1, H1, H3, K4, L1, L2, L4, N1, and O2). From a statistical point of view, all intercepts are insignificant (italic format) but show a small increase in soil sorption capacity.

Figure 4.1.2 and Table 4.1.3 show the comparison of the experimental log K_{oc} values determined in the current study with the mean value of log K_{oc} in literature for different chemical domains. It was observed that the sorption capacity of soils and sediments for phenols, organochlorines and biphenyls is higher than that of HAs, while the sorption capacity of HAs for PAHs and heterocyclic compounds is higher than that of soils and sediments. The sorption interaction of phenols (mainly chlorophenols), organochlorines and biphenyls with both mineral surfaces (aluminosilicates and metal oxides) and organic matter of soils and sediments is attributed to hydrogen bond formation (MacKay et al. 2012). The sorption interaction of PAHs on HAs is stronger than that on soils and sediments due to hydrophobic partitioning on the non-polar domain of HA. Heterocyclic compounds act as cation counter ions and are sorbed to negatively charged sites of HAs and soils and sediments via cation exchange and cation bridging, and due to the presence of -COOH, -OH, -NH_2 (negative receptors sites) in HA structure, it was expected that sorption will be little promoted (MacKay et al. 2012). The sorption capacity of HAs and soils and sediments for R-benzene (R= H, CH_3, 3(CH_3), and CHO), halobenzenes and anilines is similar. The statistical data in Table 4.1.3 shows a very high value of *rms* for such chemical domains [PAHs, R-benzene (R= H, CH_3, 3(CH_3), and CHO), halobenzenes andanilines] because of their diverse experimental data was found in literature for these compounds sorbed on various types of HAs, soils and sediments as shown in Appendix 9.16.

Table 4.1.3 Results of linear regression of exp. log K_{oc} versus exp. log K_{oc} in literature for different chemical domains.

Chemical Domain	N	r^2	q^2	rms	bias	F test	me	mne	mpe
Phenols sorbed to humic acids	7	0.09	-7.94	0.64	-0.26	5.36	0.49	-0.96	0.80
Phenols sorbed to soils & sediments	8	0.34	-0.06	0.77	-0.14	8.22	0.61	-1.11	1.10
PAHs sorbed to humic acids	6	0.67	-0.14	0.99	0.75	6.43	0.75	0.03	1.76
PAHs sorbed to soils & sediments	6	0.54	-0.47	1.13	0.83	5.03	0.87	-0.10	1.70
Organochlorines & biphenyls sorbed to humic acids	6	0.73	0.42	0.73	-0.39	11.20	0.53	-1.11	0.43
Organochlorines & biphenyls sorbed to soils & sediments	5	0.92	0.60	0.68	-0.47	13.49	0.48	-1.13	0.03
Heterocyclic compounds sorbed to humic acids	5	0.89	-0.93	0.52	-0.24	7.57	0.43	-0.59	0.49
Heterocyclic compounds sorbed to soils & sediments	6	0.70	-1.67	0.55	-0.20	8.09	0.40	-0.87	0.51
R-benzene & halobenzenes & anilines sorbed to humic acids	9	0.38	-0.18	1.13	-0.12	10.86	0.83	-1.03	2.45
R-benzene & halobenzenes & anilines sorbed to soils & sediments	10	0.39	-0.08	1.02	-0.17	12.41	0.80	-1.48	1.90

N is no. of points, r^2 is squared correlation coefficient, q^2 is predictive squared correlation coefficient, calculated as shown in Appendix 9.21 (Schüürmann et al. 2008), *rms* is root-mean-square error of correlation, *bias* is systematic error, *me* is mean error, *mne* is maximum negative error, *mpe* is maximum positive error, *F*- test was calculated at α = 0.05.

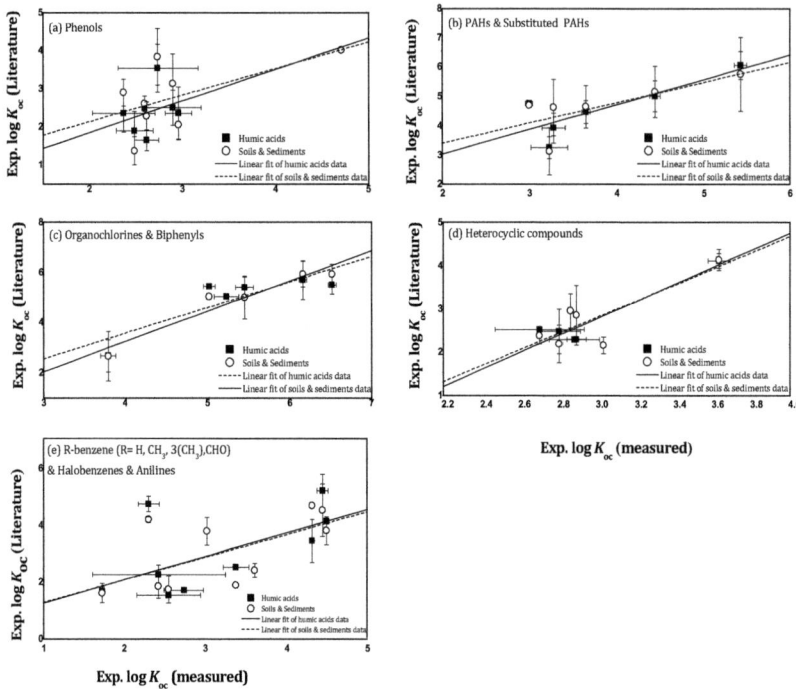

Figure 4.1.2 Exp. log K_{oc} in literature for different chemical domains versus exp. log K_{oc} in the current study.

4.1.3 Comparison with calculated log K_{oc} from different prediction methods

In the beginning of my interpretation of the results of comparisons with different prediction methods in literature, some of the data were collected from the "ChemProp" software and other data calculated from literature. The training or calibration sets for each method was checked depending on the presence of chemical domain such as PAHs, phenols, triazines, etc. as shown in Appendix 9.22. It was observed that the 2D molecular structure model (Schüürmann et al. 2006) does not include amino and nitro PAHs derivatives, pyridine derivatives, PCB 180 and PCB 202. The molecular connectivity indices model by Sabljic' et al. (1995) does not include nitro PAHs derivatives, pyridine derivatives, PCB 180 and PCB 202. The fragment constant model by Tao et al. (1999) does not include nitro PAHs derivatives, benzaldehyde, flourobenzene, and pyridine derivatives. The LSER approaches by Nguyen et al. (2005) and Poole et al. (1999) models do not include triazines, amino and nitro PAHs derivatives, pyridine derivatives, benzaldehyde, and fluorobenzene. The KOCWIN software model (calculated log K_{oc} based on log K_{ow}, the only validation set available) does not include nitro PAHs derivatives, benzaldehyde, and PCB 180 while of the model of calculated log K_{oc} based on molecular topology method does not include nitro PAHs, triazines, chlorobenzenes, PCB 180, PCB 202, benzaldehyde, and fluorobenzene. The LSER approaches by Endo et al. (2009a) and Bronner et al. (2010a) do not include amino PAHs derivatives, pyridine derivatives, triazines, benzaldehyde, fluorobenzene, PCB 180 and PCB 202. The Franco and Trapp model (2008) was applied only to unionized fractions at pH 7 with consideration of $pK_a \simeq 50$ for non-polar compounds as shown in Appendix 9.18. This model does not include PAHs and substituted PAHs, chlorobenzenes, PCBs, benzene, toluene, benzaldehyde, and fluorobenzene. The LSER approach by Neale et al. (2012) model does not include amino PAHs derivatives, pyridine derivatives, triazines except atrazine, benzaldehyde, fluorobenzene and PCB 202.

Overall, the statistical data summarized in Table 4.1.4 show that in linear regression of exp. log K_{oc} versus calc. log K_{oc}, the fragment constant model by Tao et al. (1999) is statistically robust and has the best performance compared to the results achieved with the literature methods with a low difference of 0.08 between r^2 and q^2 indicating the absence of systematic errors and due to a wide training set in the Tao model. Meanwhile the log K_{oc} of benzaldehyde could not be calculated using the Tao model. The model calculates log K_{oc} from contributions by individual atoms, bonds and functional groups that have been fitted to a data set. Obviously there is no fragment value available for an aldehyde group attached to an aromatic ring. In addition, the Tao model has the lowest root mean squared and minimum negative errors. The second best prediction performance was the 2D molecular structure model (Schüürmann et al. 2006) with a low difference between r^2 and q^2 of 0.07 indicating the absence of systematic errors and due to a wide range training set regarding the chemical domains. The LSER approach by Poole et al. (1999) with calc. descriptors has shown the third best performance with a substantial difference between r^2 and q^2 of 0.20. While with exp. input, Poole et al. (1999) model has shown a lower difference between r^2 and q^2 of 0.13.

The Franco and Trapp model (2008) has shown the forth best performance after being applied only to unionized fractions at pH 7 and also LSER approach –calc. input by Nguyen et al. (2005). Franco and Trapp model (2008) has shown a difference between r^2 and q^2 of 0.22 for unionized fractions of the current study. Recent models such as Endo et al. (2009a), Bronner et al. (2010a) and Neale et al. (2012) have shown great differences between r^2 and q^2 accompanied with higher root-mean-squared error, high minimum negative error and high positive maximum error indicating the limited quality of their original training set, which might be too narrow with regard to chemical domains in the current study and presence of systematic error. The Poole et al. (1999), Nguyen et al. (2005), and Endo et al. (2009a) (low peat concentration) exp. input LSER models have shown lower prediction performance than that of calc. input descriptors since for more complex compounds, experimental Abraham descriptors are often not available. The difference between the Poole and Nguyen models of LSER approach based on the LSER hydrogen bond donor parameter has been described in chapter 3. For the Poole model, increasing hydrogen bond acidity decreases log K_{oc}, while the opposite has been provided by the Nguyen model. Calculated log K_{oc} of KOCWIN software based on molecular topology method has shown the lowest difference between r^2 and q^2 of 0.03 indicating the high quality of its original training set and the absence of systematic errors. The relatively low regression coefficient (r^2= 0.75) of calculated log K_{oc} based on log K_{ow} by KOCWIN has been attributed to the model not fitting polar compounds whose K_{oc} values can be affected by protonation or dissociation depending on pH variation. Endo et al. (2009a) (low peat concentration) model for calculated and experimental inputs has shown the lowest prediction performance with high difference between r^2 and q^2 associated with highest rms and mpe as shown in Table 4.1.4.

Applicability domain

The current study presents the prediction performance of each method in literature for different chemical domains. Figure 4.1.3 shows root mean squared error (rms) versus prediction methods in literature for (a) PAHs, (b) phenols, (c) R-benzene (R= H, CH$_3$, 3(CH$_3$), and CHO), (d) organochlorines and biphenyls, (e) heterocyclic compounds, (f) halobenzenes and (g) anilines. The regression statistics for each chemical domain using different methods in literature are shown in Appendix 9.22.

Table 4.1.4 Results of linear regression of exp. log K_{oc} versus calc. log K_{oc} using different methods.

Model	N	r^2	q^2	rms	F-Test	bias	me	mne	mpe
2D Molecular structure[a]	61	0.80	0.73	0.59	201.48	-0.29	0.46	-1.46	1.36
KOCWIN[b] Molecular topology (top indices +fragments)	61	0.77	0.74	0.57	243.40	-0.11	0.44	-1.70	1.14
KOCWIN[b] (log K_{ow})	61	0.75	0.63	0.68	195.64	-0.30	0.53	-1.77	1.33
Molecular connectivity indices[c]	61	0.71	0.56	0.75	158.81	-0.38	0.60	-1.73	0.94
Fragment constant model[d]	60*	0.87	0.79	0.52	340.67	-0.28	0.41	-1.21	0.91
LSER (Nguyen)[e] exp.	46#	0.76	0.59	0.69	175.76	-0.13	0.51	-1.97	1.29
LSER (Nguyen)[e] calc.	61	0.81	0.57	0.73	250.19	-0.26	0.58	-1.65	1.59
LSER (Poole)[f] exp.	47#	0.80	0.67	0.61	205.30	-0.18	0.46	-1.61	1.36
LSER (Poole)[f] calc.	61	0.82	0.62	0.69	258.85	-0.32	0.57	-1.66	1.10
Franco & Trapp model (2008)[g]	52**	0.80	0.58	0.71	108.03	-0.51	0.57	-1.79	0.53
LSER (Endo 2009 – low Peat)[h] exp.	61	0.72	-1.20	1.66	112.57	1.16	1.33	-1.63	4.89
LSER (Endo 2009 – low Peat)[i] calc.	61	0.83	-1.67	1.83	118.79	1.25	1.46	-1.39	4.83
LSER (Endo 2009 – high Peat)[j] exp.	61	0.80	0.15	1.04	194.47	0.27	0.81	-2.26	2.47
LSER (Endo 2009 – high Peat)[k] calc.	61	0.79	-0.17	1.21	176.37	0.01	0.99	-2.09	3.00
LSER (Bronner 2010a)[l] exp.	61	0.79	0.34	0.91	213.97	-0.22	0.72	-2.34	2.08
LSER (Bronner 2010a)[m] calc.	61	0.74	-0.31	1.29	147.11	-0.62	1.01	-2.86	1.80
LSER (Neale 2012)[n] exp.	61	0.78	-0.35	1.31	152.12	0.64	1.05	-2.17	3.20
LSER (Neale 2012)[o] calc.	61	0.77	-0.27	1.26	165.03	0.26	1.05	-1.91	3.06

a) Schüürmann et al. 2006. b) KOCWIN software. c) Sabljić et al. 1995.d) Tao et al. 1999. e) Nguyen et al. 2005. f) Poole et al. 1999.g) Franco et al. 2008. h-k) Endo et al. 2009a. l-m) Bronner et al. 2010a. n-o) Neale et al. 2012. #Experimental Abraham descriptors were not available for all solutes (see Appendix 9.13, 9.14 and 9.17). *Log K_{oc} of benzaldahyde could not be calculated. ** The model was used only for 52 unionized fractions at pH 7 ± 0.20. N is no. of points, r^2 is squared correlation coefficient, q^2 is predictive squared correlation coefficient, calculated as shown in Appendix 9.21 (Schüürmann et al. 2008), rms is root-mean-square error of correlation, bias is systematic error, me is mean error, mne is maximum negative error, mpe is maximum positive error, F-test was calculated at $\alpha = 0.05$.

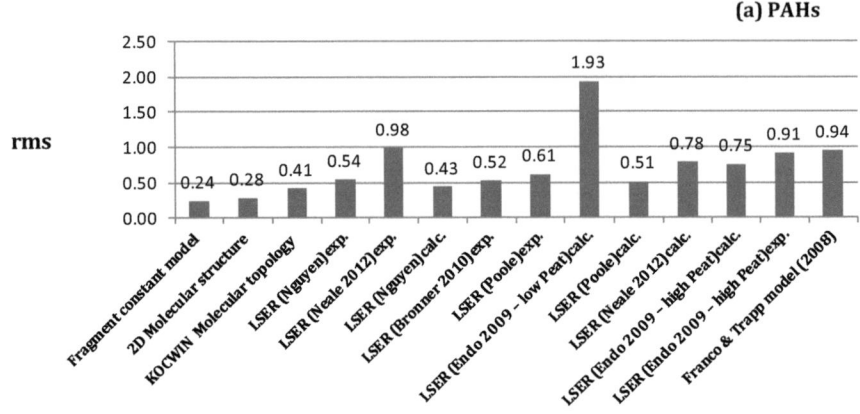

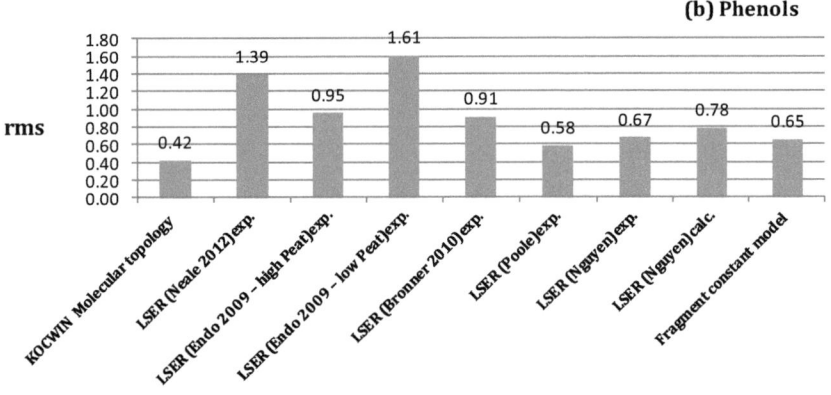

Figure 4.1.3 Root mean squared error (*rms*) versus prediction methods in literature for different chemical domains. (a) PAHs, (b) Phenols.

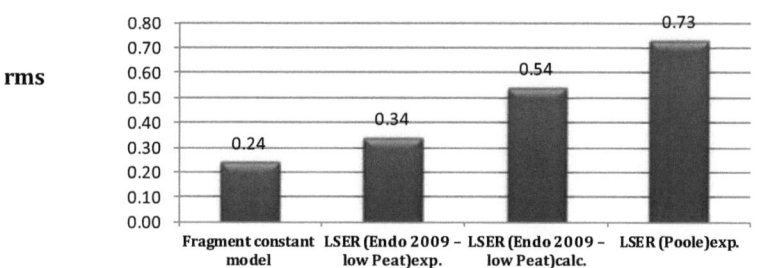

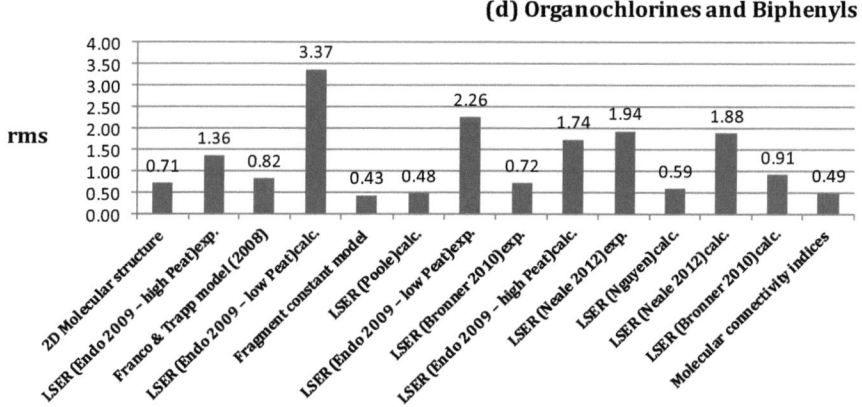

Figure 4.1.3 Continued. (c) R-benzene (R= H, CH$_3$, 3(CH$_3$), and CHO), (d) Organochlorines and Biphenyls.

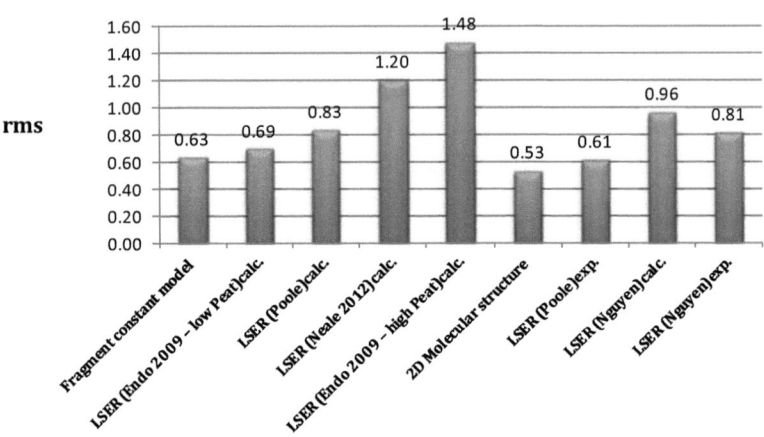

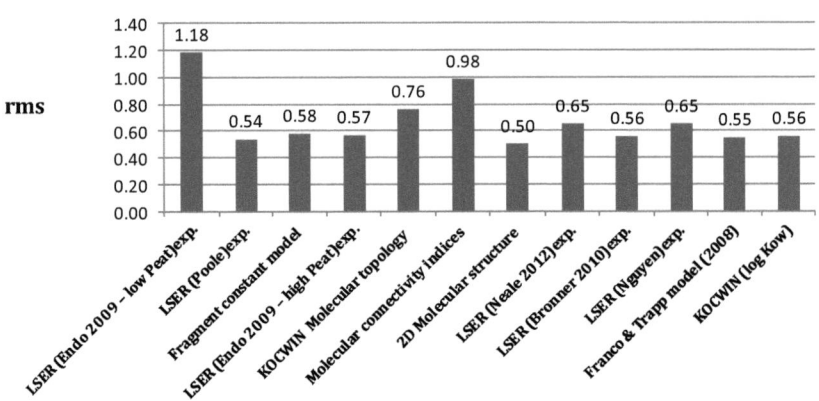

Figure 4.1.3 Continued. (e) Heterocyclic compounds, (f) Halobenzenes.

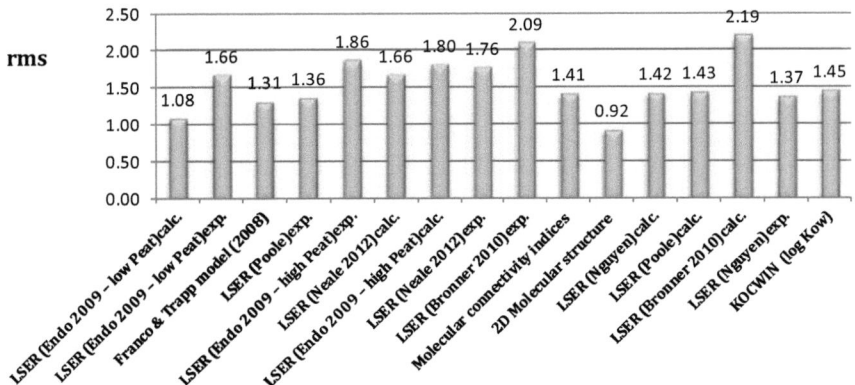

Figure 4.1.3 Continued. (g) Anilines.

For PAHs, it was found that the fragment constant model by Tao et al. (1999) (r^2= 0.92, q^2= 0.92, rms = 0.24) and the 2D molecular structure model by Schüürmann et al. (2006) (r^2= 0.90, q^2= 0.89, rms = 0.28) show the best performance of regression. The worst performance of prediction for PAHs has been shown by the LSER approaches by Endo et al. (2009a) low peat concentration – calc. input (r^2 = 0.67, q^2 = -4.12, rms = 1.93) and Neale at al. (2012) – exp. input (r^2 = 0.80, q^2 = -0.33, rms = 0.98).

For phenols, it was found that KOCWIN software based on molecular topology method (r^2 = 0.86, q^2 = 0.75, rms = 0.42) and LSER approach by Poole et al. (1999)–exp. input (r^2 = 0.58, q^2 = 0.38, rms = 0.58) have the best performance, while the worst performance of prediction for phenols has been shown by the LSER approaches by Endo et al. (2009a) low peat concentration – exp. input (r^2 = 0.69, q^2 = -2.81, rms = 1.61) and Neale at al. (2012)–exp. input (r^2 = 0.70, q^2 = -1.83, rms = 1.39).

For R-benzene (R= H, CH_3, $3(CH_3)$, and CHO), it was found that the fragment constant model by Tao et al. (1999) for only 3 points of data ($r^2 = 1$, $q^2 = 0.92$, $rms = 0.24$) and the Endo et al. (2009a) low peat concentration – exp. input model for only 4 points of data ($r^2 = 0.78$, $q^2 = 0.76$, $rms = 0.34$) have the best performance of regression as shown in Appendix 9.22.

For organochlorines and biphenyls, it was observed that the fragment constant model by Tao et al. (1999) ($r^2 = 0.86$, $q^2 = 0.73$, $rms = 0.43$), the LSER approach by Poole et al. (1999) – calc. input ($r^2 = 0.85$, $q^2 = 0.67$, $rms = 0.48$) and the molecular connectivity indices model by Sabljić et al. (1995) ($r^2 = 0.75$, $q^2 = 0.65$, $rms = 0.49$) have the best performance of regression (good prediction), while the Endo et al. (2009a) models of low peat concentration for both calculated and experimental input have the worst performance (for calc. input $r^2 = 0.87$, $q^2 = -15.50$, $rms = 3.37$ and for exp. input $r^2 = 0.84$, $q^2 = -6.43$, $rms = 2.26$) as shown in Appendix 9.22.

For heterocyclic compounds, it was observed that the 2D molecular structure model by Schüürmann et al. (2006) ($r^2 = 0.61$, $q^2 = 0.02$, $rms = 0.53$), and the LSER approach by Poole et al. (1999) – exp. input ($r^2 = 0.59$, $q^2 = -0.12$, $rms = 0.61$) have the best performance while the Endo et al. (2009a) model of high peat concentration – calc. input ($r^2 = 0.62$, $q^2 = -6.69$, $rms = 1.48$) and the Neale et al. (2012) – calc. input model ($r^2 = 0.63$, $q^2 = -4.06$, $rms = 1.20$) have the worst performance as shown in Appendix 9.22.

For five points of halobenzenes, it was found that the 2D molecular structure model by Schüürmann et al. (2006) ($r^2 = 0.83$, $q^2 = 0.75$, $rms = 0.50$), and the LSER approach of Poole et al. (1999) – exp. input ($r^2 = 0.84$, $q^2 = 0.70$, $rms = 0.54$) have the best performance of regression while the Endo et al. (2009a) model of low peat concentration – exp. input ($r^2 = 0.86$, $q^2 = -0.39$, $rms = 1.18$) and the molecular connectivity indices model by Sabljić et al. (1995) ($r^2 = 0.84$, $q^2 = 0.04$, $rms = 0.98$) have the worst performance.

For three points of anilines, it was found that no perfect model can predict the log K_{oc} of heterocyclic compounds. The 2D molecular structure model by Schüürmann et al. (2006) ($r^2 = 0.63$, $q^2 = -0.83$, $rms = 0.92$) has the best performance of regression while the LSER approaches by the Bronner et al. (2010a) model for both calculated and experimental input have the worst performance (for calc. input $r^2 = 0.62$, $q^2 = -9.42$, $rms = 2.19$ and for exp. input $r^2 = 0.65$, $q^2 = -8.49$, $rms = 2.09$) as shown in Appendix 9.22.

One of the main findings of these statistical analyses was the good performance of the Franco and Trapp model (2008) for both halobenzenes ($r^2 = 0.81$, $q^2 = 0.70$, $rms = 0.55$) and organochlorines and biphenyls ($r^2 = 0.88$, $q^2 = 0.03$, $rms = 0.82$) by using equation (9.2) of the calculation of log K_{oc} for weak acids as shown in Appendix 9.18.

Figure 4.1.4 shows the behavior of each chemical domain for the used prediction methods. The best performance was found for the fragment constant model by Tao et al. (1999) for most of chemical domains in the current study, while the 2D molecular structure model by Schüürmann et al. (2006) has shown an overestimation for some of phenols and underestimation for some organochlorines and biphenyls. The LSER approach by Poole et al. (1999) – exp. input has shown overestimation for some PAHs and underestimation for anilines. The calculated log K_{oc} data by KOCWIN software based on molecular topology method has shown underestimation for anilines and some heterocyclic compounds.

The molecular connectivity indices model by Sabljić et al. (1995) has shown underestimation for halobenzenes, anilines, some heterocyclic compounds and PAHs. The LSER approach by Nguyen et al. (2005) – calc. input has shown underestimation of phenols, anilines and heterocyclic compounds. The LSER approaches by Endo et al. (2009a) of both low peat – calc. input and high peat – exp. input have shown overestimation for halobenzenes, organochlorines, biphenyls, phenols and PAHs. The LSER approach by Neale et al. (2012) has shown overestimation for organochlorines, biphenyls, phenols and some PAHs and underestimation for anilines and some triazines. Franco and Trapp model (2008) have shown underestimation for organochlorines, biphenyls, some PAHs and triazines.

After consideration for the small training set in the current study and the comparisons with different existing methods, it was concluded that the performance of the methods based on the molecular structure and fragments by the contributions of individual atoms, bonds and functional groups is better than LSER methods for exp. and calc. input in log K_{oc} prediction. Therefore, these methods were recommended to predict the sorption coefficients of non-polar compounds but still no existing methods well fitted for polar compounds, it will be discussed in more details in chapter 4-2.

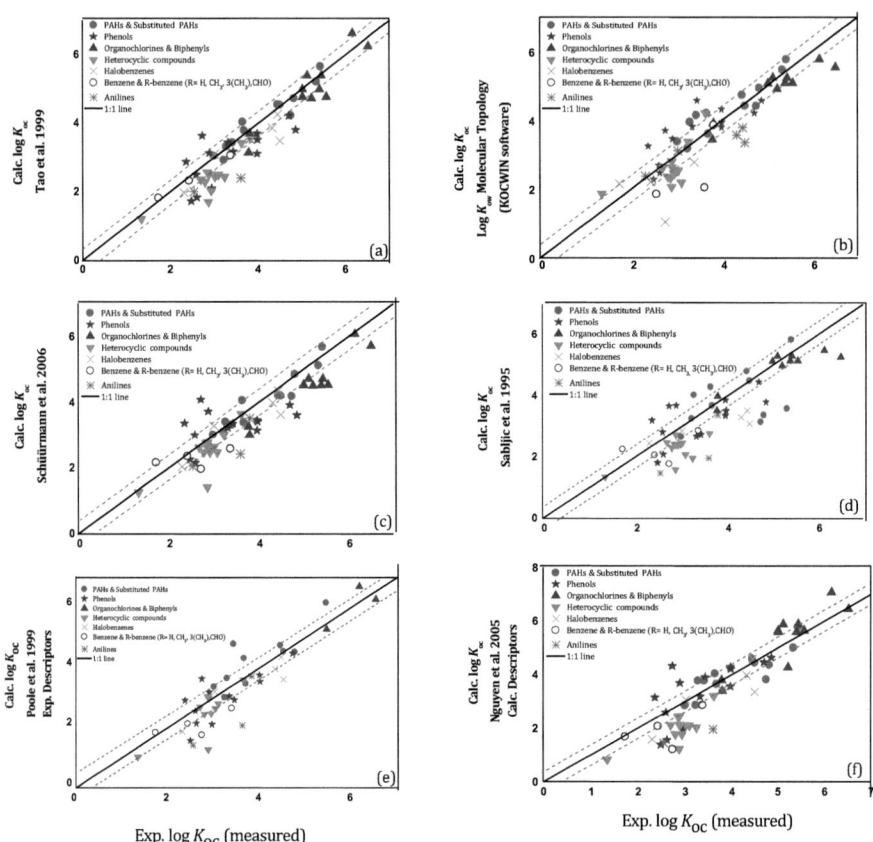

Figure 4.1.4 Calc. log K_{oc} in literature for different chemical domains versus exp. log K_{oc} measured.

(a) Fragment constant method (Tao et al. 1999)

(b) KOCWIN software (log K_{ow}) – molecular topology

(c) 2D Molecular structure (Schüürmann et al. 2006)

(d) Molecular connectivity indices (Sabljić et al. 1995)

(e) LSER $_{exp.\ input}$ (Poole et al. 1999)

(f) LSER $_{calc.\ input}$ (Nguyen et al. 2005)

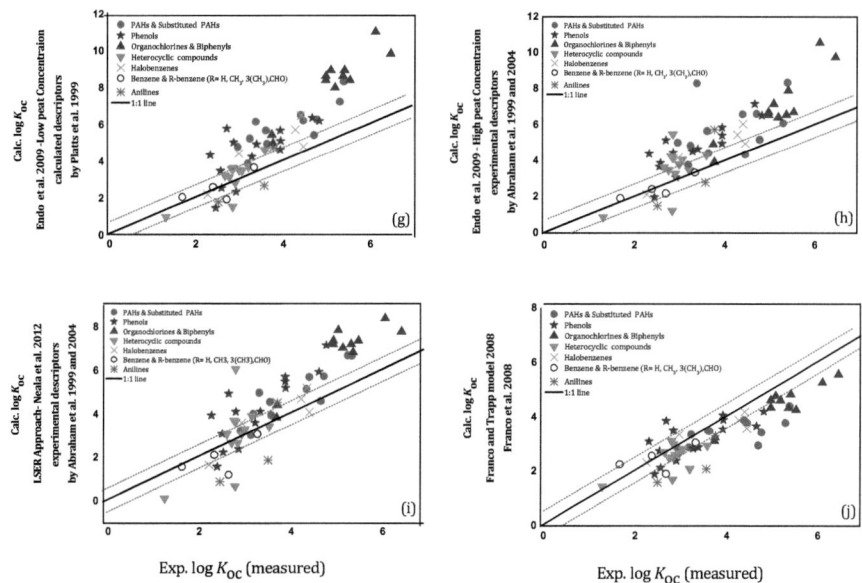

Figure 4.1.4 Continued.

(g) LSER $_{calc.\ input}$ (Endo et al. 2009a – low Peat conc.)

(h) LSER $_{exp.\ input}$ (Endo et al. 2009a – high Peat conc.)

(i) LSER $_{exp.\ input}$ (Neale et al. 2012- Aldrich HA in literature)

(j) Franco and Trapp model for unionized fractions at pH7 (Franco et al. 2008)

4.1.4 Refitting of the common K_{ow}-approach for sorption to humic acid

It should be kept in mind that the soil organic carbon content seems to be mainly responsible for neutral organic chemicals sorption, which leads to the assumption that the sorption process onto soil can be considered as a partitioning process between water and a lipophilic soil phase. One of the prediction models for soil sorption, the classical approach, has used octanol as surrogate for the organic carbon in soils, sediments and organic matter (Gawlik et al. 1997 and Schüürmann et al. 2007).

Figure 4.1.5 shows that log K_{ow} – log K_{oc} correlation for sixty-one substances that have different chemical structures with a wide range of hydrophobicity in terms of log K_{ow} (0.65 – 8.27). There is no prefect fitting of log K_{oc} for the dataset used due to presence of different chemical classes with different chemical structures such as phenols, heterocyclic, anilines, organochlorines and PAHs (r^2 = 0.75, rms = 0.56, q^2_{cv} = 0.74 and rms_{cv} = 0.58) as shown in Table 4.1.5. It shows also that heterocyclic compounds, phenols and anilines do not fit well with the K_{ow} approach to predict K_{oc}, and most of their points are outside the 95 % confidence limit of the linear fit which may be attributed to their hydrophobicities being an insufficient descriptor for the soil adsorption process.

The log K_{ow} – log K_{oc} correlations (Karickhoff et al. 1981, Sabljić et al. 1995 and Gawlik et al. 1997) at pH 7 ± 0.2 and 25 ± 2 °C have been performed and their linear regression results with full statistics have been calculated and illustrated in Table 4.1.5 with comparison to literature. The experimental dataset of log K_{ow} has been used in modelling and in cases where no experimental data is available, the calculated log K_{ow} has been used with the help of KOWWIN software.

For different chemical classes, significantly different intercepts, slopes and regression coefficients have been found in Table 4.1.5. For non-ionogenic compounds (organochlorines, biphenyls and PAHs), their hydrophobicity properties contribute to the sorption process. This concurs with the review of existing literature which has shown that the K_{ow} approach can be used for only certain chemical classes such as organchlorines, PAHs, biphenyls and difficult for a large heterogeneous neutral hydrophobic dataset with chemically different structures where sorption is governed by unspecific dispersion interactions (Gawlik et al. 1997 and Doucette 2003, Nguyen et al. 2005). The classical model implies that the absorption into organic matter is the dominant process where the sorption behavior is mainly caused by van-der-Waals forces. For neutral acids as phenols and neutral bases as heterocyclic compounds which have hydroxyl and amino groups can participate in hydrogen bonding, the K_{ow} approach does not fit well as the K_{oc} predictor attributed to adsorption phenomena such as the association of ionized chemicals at polar mineral surfaces is not considered in that approach. Log K_{ow} has limited scope to predict log K_{oc} with increasing complexity of the molecular structure through increasing functional groups (Karickhoff et al. 1981, Sabljić et al. 1995 and Gawlik et al. 1997), and in particular, hydrogen bond sites, therefore the applicability of K_{ow} to predict K_{oc} will decrease. Some investigations in literature agree with the result that the K_{ow} approach is not a reliable model to predict K_{oc} for polar compounds. For example, von Oepen et al. (1991a) have mentioned that K_{ow} as a predictor for K_{oc} is not reliable for compounds which have mostly carboxylic or amino groups. Pussemier et al. (1989) have observed that the K_{ow} approach is not reliable with increasing polarity, resulting in the underestimation of log K_{oc} for acids and in the overestimation for bases. The proposed models of using the classical approach of K_{ow} to predict K_{oc} for specific chemical classes have been compared with literature models. It was found that Sabljić et al. (1995) have two different models for hydrophobic and non-hydrophobic compounds, the contribution of hydrophobicity property in the sorption process being decreased with decreasing hydropobicity of chemicals.

Log K_{oc} – log K_{ow} correlation for anilines, benzene and substituted benzene in the present study are insignificant, therefore no proposed prediction model is fitting this chemical class. It was found an agreement between the proposed model in the current study for organochlorines and biphenyls and the establish model by Chin et al. (1988) for some of the organochlorines. A slight difference has been found between the proposed models for PAHs and substituted PAHs and the existing model by Gerstl (1990) due to three compounds having amino groups as substitution in PAHs, which lowers the contribution of hydrophobicity and the ability to form hydrogen bonds due to the presence of hydrogen bond donors. For organochlorine and PAHs, the absorption into OM is the dominant process where the sorption behavior is mainly caused by van-der-Waals forces.

For phenols and heterocyclic compounds, the results have shown an agreement with the existing model by Sabljić et al. (1995), indicating a weak contribution of the hydrophobicity property in the sorption process, which can be attributed to the fact that the hydroxyl group of phenols and the electron pair of the nitrogen atoms of heterocyclic compounds can be used as hydrogen bond acceptors, so the sorption can occur via non-dispersive interactions. This result agrees with literature. Feng et al. (1996) have shown that the difficulty to apply the log K_{ow} – log K_{oc} correlation for polar compounds can be attributed to the fact that the organic carbon phase is mainly responsible for adsorption to soil where there are more cohesive and stronger hydrogen bond donor solvents than n-octanol. The sorption process of polar compounds depends on non-hydrophobic and non-dispersive interactions. Schüürmann et al. (2007) have proposed a model for large heterogeneous neutral organic chemical dataset including different types of chemical domains as shown in Table 4.1.5. This agrees with the proposed model for another different dataset which includes nine polar compounds that are protonated or dissociated at pH 7 ± 0.2 depending on their pK_a values, therefore the regression correlation and prediction performance will be improved by exclusion of the ionized fractions at pH7 ± 0.2 (J2, J3, J4, K1, K2, M1, M2, M3, and M4 as shown in Scheme 3.1) and by remodelling for unionized fractions.

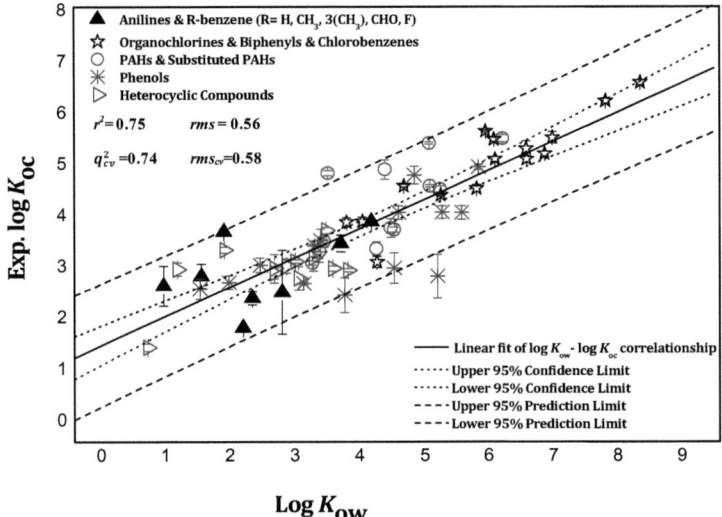

Figure 4.1.5 Exp. log K_{oc} versus log K_{ow} in neutral water for different chemical classes. (see Table 4.1.5)

Table 4.1.5 Results of linear regression of log K_{oc} versus log K_{ow} and comparison with similar models in literature.

Substances	Model Equation	N	r^2	rms	q_{cv}^2	rms_{cv}	F-Test	bias	me	mne	mpe
Non-ionogenic compounds	log K_{oc} = (0.71±0.04) log K_{ow} +(0.69±0.21)	28	0.92	0.32	0.92	0.35	318.14	0.00	0.26	-0.69	0.67
Sabljić et al.1995 Hydrophobics	log K_{oc} =0.81 log K_{ow} + 0.10	81	0.89	-	-	-	629	-	-	-	-
Unionized fractions of acids mostly phenols at pH7	log K_{oc} = (0.43±0.05) log K_{ow} + (1.97±0.20)	9	0.91	0.21	0.87	0.31	73.66	0.00	0.16	-0.44	0.35
Sabljić et al. 1995 Nonhydrophobics	log K_{oc} =0.52 log K_{ow} + 1.02	390	0.63	-	-	-	666	-	-	-	-
Unionized fractions of bases at pH7	log K_{oc} = (0.54±0.17) log K_{ow} + (1.77±0.56)	15	0.42	0.64	0.30	0.78	9.43	0.00	0.58	-1.12	0.93
Total unionized fractions	log K_{oc} = (0.58±0.04) log K_{ow} + (1.45±0.18)	52	0.80	0.48	0.79	0.51	201.3	0.00	0.36	-1.31	0.95
All compounds at pH7	log K_{oc} = (0.57±0.04) log K_{ow} + (1.43±0.19)	61	0.75	0.56	0.74	0.58	174.65	0.00	0.42	-1.36	1.61
Schüürmann et al. 2007 General Model	Log K_{oc} = 0.71 log K_{ow} +0.62	167	0.87	0.52	0.87	-	-	-	-	-	-
Organochlorines & biphenyls	log K_{oc} = (0.63±0.08) log K_{ow} + (1.17±0.46)	15	0.84	0.36	0.82	0.42	68.42	0.00	0.28	-0.67	0.81
Chin et al. 1988 CCl4, Lindane, TCBs, PCBs, Chlordane	log K_{oc} = 0.61 log K_{ow} +1.18	5	0.94	-	-	-	-	-	-	-	-
PAHs & substituted PAHs	log K_{oc} = (0.67±0.21) log K_{ow}	12	0.50	0.57	0.44	0.69	10.21	0.00	0.46	-1.22	0.75
Gerstl 1990 PAHs	log K_{oc} = 0.76 log K_{ow} +1.05	20	0.95	-	-	-	-	-	-	-	-
Phenols	log K_{oc} = (0.42±0.13) log K_{ow} + (1.73±0.52)	14	0.48	0.57	0.41	0.32	10.96	0.00	0.46	-0.99	1.13
Sabljić et al. 1995 Phenols & benzonitriles	log K_{oc} = 0.57 log K_{ow} +1.08	24	0.74	-	-	-	65	-	-	-	-
Heterocyclic compounds mostly triazines	log K_{oc} = (0.34±0.14) log K_{ow} + (1.97±0.39)	12	0.37	0.41	-0.19	0.64	5.88	0.00	0.31	-0.65	0.84
Sabljić et al. 1995 Triazines	log K_{oc} = 0.30 log K_{ow} + 1.50	16	0.27	-	-	-	6.6	-	-	-	-

N is no. of points, r^2 is squared correlation coefficient, q_{cv}^2 is predictive squared correlation coefficient using leave-one-out- cross validation, calculated as shown in Appendix 9.21 (Schüürmann et al. 2008), rms is root-mean-square error of correlation, rms_{cv} is root-mean-square error of prediction using leave-one-out cross validation, bias is systematic error, me is mean error, mne is maximum negative error, mpe is maximum positive error, F-test was calculated at α = 0.05, see all formulas in Appendix 9.21. The unionized fractions means non-ionogenic compounds and unionized fractions of acids and bases at pH 7 of (1> f_u > 0.995).

4.1.5 Refitting of LSER-approach for sorption to humic acid

In the beginning of the application of the LSER-approach, the cross-correlation for both experimental and calculated descriptors of the investigated solutes has been checked and shown in Appendix 9.15. The solvent parameters have been calculated using multiple regressions at 95 % confidence limit. The experimental solute descriptors of Abraham and co-workers (Abraham et al. 1987, 1991, 1994, 1999 and 2004) and calculated solute descriptors by increment method (Platts et al. 1999) have been used in the regressions. Tables 4.1.6 and 4.1.7 show some of the findings that have been observed and will be discussed in detail in the following points:

Firstly, when applying the LSER approach with experimental and calculated input descriptors for neutral solutes, it has been found that the +v term, the energy difference of cavity formation between water phase and HA phase, is significant in the sorption process. It indicates that the neutral solutes prefer to be sorbed to HA due cavity formation energy in HA phase lower than water. The multiple regressions of LSER exp. input descriptors showed that both s, b are negative. Water is more polar than HA, which indicates that solutes prefer water as a stronger hydrogen bond donor than HA. LSER exp. input showed a better performance than LSER calc. input.

Secondly, when applying the LSER approach with experimental and calculated input descriptors for acids, it has been found that the v term, the energy difference of cavity formation, is significant in the sorption process for calculated descriptors while all solvent parameters are insignificant at confidence level 95% for experimental descriptors (some data still needed). This indicates that acids prefer to be sorbed to HA with lower cavity formation energy in HA phase due to the cavity formation energy in water phase (penalty) will destroy the aqueous hydrogen bond but in HA phase the penalty compensated and become less participation in hydrogen bonding . This behavior was observed for acids with smaller positive value of v term. LSER exp. input has shown a great difference between r^2 and q_{cv}^2 performance than LSER calc. input, due to the fact that Abraham experimental descriptors are not available for all solutes and the limited dataset.

Thirdly, when applying the LSER approach with experimental and calculated input descriptors for bases, it has been found that the b term is significant in the sorption process for calculated descriptors while all solvent parameters are insignificant at confidence limit 95 % for experimental descriptors. It was found that bases prefer water molecules as hydrogen bond donor more than the carboxylic groups of HA. Thus, the sorption can occur via hydrogen bonding. LSER exp. input has shown a great difference between r^2 and q_{cv}^2 performance than LSER calc. input, due to the fact that Abraham experimental descriptors are not available for all solutes and the narrow dataset.

Fourthly, when applying the LSER approach with experimental and calculated input descriptors to ionized compounds (acids and bases ionized at pH 7), it has been found that all solvent parameters are insignificant at confidence limit 95% for experimental and calculated descriptors. LSER exp. input has shown a great difference between r^2 and q_{cv}^2 performance than LSER calc. input. Therefore, this approach failed in the prediction of ionized compounds.

Fifthly, when applying the LSER approach with experimental and calculated input descriptors to all compounds at pH 7, it has been found that the *b* and *v* terms are significant in the sorption process. Therefore, this result proved that organic solutes prefer to be sorbed to HA via hydrophobic partitioning and hydrogen bonding. The sorption process can occur by non-specific intermolecular interactions such as van-der-Waals force and by specific interactions such as hydrogen bond formation. Moreover, water molecules are more polar than HA, therefore water is a better hydrogen bond donor than the carboxylic, phenolic and amino groups of HA which play a role as hydrogen bond acceptors.

For total sixty-one solutes, LSER exp. input has shown a larger difference between r^2 and q_{cv}^2 performance than LSER calc. input. By exclusion of the nine ionized fractions at pH 7 and remodelling again for fifty-two unionized fractions, it has been found that the same results of *b* and *v* are significant but with a slight increase of the negativity of *b* term, indicating that water molecules will be more favorable hydrogen bond donors than HA carboxylic groups. The prediction performance has been improved after the exclusion of ionized compound.

Sixthly, in comparison with data in literature to investigate the sorption mechanisms relevant to the two different types of sorbents, HA on one hand and soils and sediments on the other hand, a difference between LSER exp. input and LSER calc. input has been observed. The first (LSER exp. input) showed that the sorption to HA, soils and sediments can occur via specific (*a, b*) and non-specific interactions (*e, v*) due to the avoidance of high energy (penalty) associated with cavity formation. While the second (LSER calc. input) shows that the energy difference of cavity formation between two phases *v* term is insignificant at confidence limit 95% for both HA and soil data. It shows the sorption can occur via specific interactions (*a, b*) through hydrogen bond formation only. Moreover, LSER exp. input shows sorption to soils and sediments has high energy difference of cavity formation between water and soil than that between water and HA. In addition, the mineral surface of soils and sediments can participate in the sorption process through surface complexation via ligand exchange, cation exchange, anion exchange, cation bridging and electron donor-acceptor (EDA) interactions (MacKay et al. 2012).

The results of linear regressions of the modelling of different chemical classes are shown in Table 4.1.8. Three types of QSAR modelling have been used: log K_{ow} approach, LSER calc. input descriptors and a combination of both (log K_{ow} + LSER). Anilines and R-benzene (R= H, CH_3, 3(CH_3), CHO, F) show a good fitting with E, V solute parameters of LSER which indicates that only the difference in energy penalty of cavity formation between water and HA drives sorption to increase or decrease.

Organochlorines and PAHs have a good fitting with the K_{ow} approach, which implies that the hydrophobicity property is a good descriptor to predict K_{oc} for these compounds. S, the polarizability of solute, can be used to improve the prediction of sorption for organochlorine and PAHs, which has also been described by Zhu et al. (2005b). A slight improvement of correlation coefficients of organochlorine by 0.08 has been shown by using LSER calc. input, the polarizability is more dominant in the sorption process due to the inductive effect of chlorines atoms and their high electronegativity and strong electron withdrawing, where the associated π structures become electron deficient and thus act as π-electron acceptors to interact strongly with π-electron donor of polarized π- electron rich HSs (Liu et al. 2008 and Qu et al. 2008).

It has been found that the sorption of PAHs can be enhanced by the presence of aromatic components in OM, and an increase in the polarizability of the PAHs in aromatic-rich HSs has been attributed to the preferential interaction of PAHs with aromatic moieties of SOM. It was observed that the prediction of the K_{oc} of PAHs onto HA has been improved as r^2 increased by 0.28 from the common log K_{ow} approach (log K_{oc} = a log K_{ow} + b) to the new proposed model (log K_{oc} = a log K_{ow} + S_{est} + c) as shown in Table 4.1.8. This can be explained by the fact that the higher polarizability of aromatic-rich HSs can increase nonspecific interactions such as van-der-Waals interactions between PAHs and sorbent due to the formation of charge transfer complexes (π-π interactions) (Zhu et al. 2004, Liu et al. 2008 and Qu et al. 2008) in which the PAHs act as electron donors and the aromatic moieties of HA act as electron acceptors.

The K_{ow} approach alone is insufficient to predict the log K_{oc} of polar compounds which can be protonated or deprotonated in neutral water systems depending on their pK_a values. Therefore, low correlation coefficients between log K_{ow} and log K_{oc} have been found for both heterocyclic compounds and phenols, attributed to protonation or deprotonation processes at pH 7 ± 0.2. As mentioned in published literature (Bi et al. 2007), the sorption of ionizable N-heterocyclic compounds is dominated by cation exchange and surface complex formation rather than partitioning into SOM controlling their overall sorption. When applying the LSER approach to Heterocyclic compounds, an insignificant correlation coefficient of sorption was found, and that approach failed in the prediction of reliable values of log K_{oc} for N-heterocyclic compounds. This fact agrees with Endo et al. (2008), who mentioned that heterocyclic aromatic compounds appear to undergo additional interactions that are still not accounted for by the LSER approach. These additional interactions considerably enhance both sorption capacity and nonlinearity.

The log K_{oc} prediction of phenols has been improved by the addition of A_{est} (hydrogen bond donor parameter of the solute as one of the specific interaction terms of LSER calculated descriptors) to log K_{ow} in a new combination between two different approaches. It has been shown that the correlation coefficient r^2 was improved by 0.18, indicating that the sorption of phenols can occur through hydrogen bond formation. An additional greatest improvement by 0.40 was observed when only LSER calc. input was applied. When applying LSER approach calc. input descriptors to phenols, it has been found that the sorption of phenols can occur through specific and non-specific interactions. *a, v* are significant, which implies that van-der-Waals forces with hydrogen bond formation are the dominant mechanisms of sorption in addition to the water molecules becoming more favorable to be hydrogen bond acceptors than HA (HA is considered a weak hydrogen bond acceptor).

It has been observed that the main findings in the present study agree with the Abraham LSER calibrated model by Poole et al. (1999), Bronner et al. (2010a) and Neale et al. (2012) as shown in the following models:

Poole et al. (1999) have proposed equation (4.1.1)

$$\log K_{oc} = 0.74\, E - 0.31\, A - 2.27\, B_0 + 2.09\, V + 0.21 \quad (4.1.1)$$
$$r^2 = 0.955,\, n = 131,\, rms = 0.248,\, F = 655$$

Bronner et al. (2010a) have proposed equation (4.1.2)

$$\log K_{oc} = 0.81 \pm (0.08)\, E - 0.61 \pm (0.11)\, S - 0.21 \pm (0.14)\, A - 3.44 \pm (0.18)\, B + 2.99 \pm (0.11)\, V$$
$$-0.29 \pm (0.12) \qquad r^2 = 0.921,\, n = 79,\, SE = 0.25 \quad (4.1.2)$$

Neale et al. (2012) have proposed equation (4.1.3)

$$\log K_{DOC-w} = 0.59 \pm (0.08)\ E - 0.52 \pm (0.12)\ S + 0.63 \pm (0.19)\ A - 3.40 \pm (0.16)\ B + 3.94 \pm (0.16)\ V$$
$$-0.85 \pm (0.16) \qquad r^2 = 0.971,\ n = 52,\ SD = 0.29 \qquad (4.1.3)$$

where the sorption can occur via specific and non-specific intermolecular interactions.

Table 4.1.6 The multilinear regression statistics of phase parameters using LSER (experimental descriptors by Abraham et al. 1987, 1994a, 1994b, 1999 and 2004).

Substances	e	s	a	b	v	c	r^2	rms	q_{cv}^2	rms_{cv}	N	F-Test
Non-ionogenic compounds	0.46 ±0.22	-0.89 ±0.29	2.71 ±6.56	-2.20 ±0.64	2.93 ±0.32	0.47 ±0.28	0.94#	0.30#	0.89#	0.44#	20	43.71
Acids	-0.57 ±0.68	1.12 ±0.99	-0.22 ±0.20	-1.35 ±0.41	1.45 ±0.14	1.36 ±0.45	0.99	0.04	-4.39	1.58	7	31.58
Bases	-0.05 ±0.65	1.61 ±1.30	-0.81 ±1.82	-1.30 ±0.80	0.17 ±0.44	2.01 ±1.30	0.93	0.16	-1.27	1.06	11	13.03
Ionized compounds	-0.62 ±0.93	0.25 ±2.68	0.57 ±4.60	2.90 ±7.44	3.06 ±1.26	-1.32 ±1.81	0.87	0.30	-8.12	3.10	9	4.22
Unionized fractions	0.63 ±0.21	-0.36 ±0.30	0.42 ±0.33	-2.02 ±0.26	1.80 ±0.24	1.22 ±0.26	0.85	0.40	0.73	0.55	38	36.37
All compounds at pH 7	0.48 ±0.22	-0.04 ±0.30	-0.28 ±0.28	-1.66 ±0.27	1.68 ±0.24	1.15 ±0.27	0.80	0.47	0.70	0.59	47	32.78
Humic acids in literature	0.83 ±0.36	0.08 ±0.60	-1.43 ±0.51	-2.25 ±0.47	1.03 ±0.48	1.69 ±0.43	0.81	0.59	0.74	0.72	31	21.60
Soils & Sediments in literature	0.80 ±0.32	0.28 ±0.53	-0.98 ±0.44	-2.40 ±0.37	1.19 ±0.38	1.31 ±0.36	0.84	0.52	0.78	0.64	34	30.36

The coefficients e (solvent dispersion interaction), s (the ability of solvent phase to undergo dipole-dipole and dipole-induced dipole interactions with a solute), a (the complementary solvent hydrogen bond basicity), b (the solvent phase hydrogen bond acidity), v (endoergic cavity effect + exoergic solute-solvent effect), and the constant c (solvent specific free energy contribution), are calculated from solutes descriptors via multiparameter linear regression analysis of the dataset. N is no. of points, r^2 is squared correlation coefficient, q_{cv}^2 is predictive squared correlation coefficient using leave-one-out -cross validation, calculated as shown in Appendix 9.21 (Schüürmann et al. 2008), rms is root-mean-square error of correlation, rms_{cv} is root-mean-square error of prediction using leave-one-out cross validation, F-test was calculated at $\alpha = 0.05$. Italic format indicates insignificant values at 95% confidence interval. # means A parameter = zero for nineteen compounds, therefore it was excluded from statistics calculation. Ionized compounds means ionized acids and bases at pH 7 of $f_u < 0.995$ for acids and bases as shown in chapter 2, unionized fractions means non-ionogenic compounds and unionized fractions of both acids and bases at pH 7 of $1 > f_u > 0.995$.

Table 4.1.7 The multilinear regression statistics of phase parameters using LSER (calculated descriptors by Platts et al. 1999).

Substances	e	s	a	b	v	c	r^2	rms	q_{cv}^2	rms_{cv}	N	F-Test
Non-ionogenic compounds	0.30 ±0.48	0.06 ±0.61	-0.81 ±3.91	-1.84 ±0.92	2.06 ±0.44	0.72 ±0.27	0.91	0.35	0.84	0.50	28	44.39
Acids	-0.83 ±1.09	1.17 ±1.35	0.11 ±0.45	-2.53 ±1.26	1.65 ±0.27	1.92 ±0.63	0.95	0.16	0.57	0.55	9	10.98
Bases	0.89 ±0.65	0.94 ±0.90	0.84 ±0.83	-2.48 ±0.51	0.56 ±0.75	1.33 ±0.63	0.95	0.19	0.86	0.35	15	33.13
Ionized compounds	-3.79 ±4.55	2.08 ±5.31	-0.58 ±1.96	0.09 ±5.50	5.57 ±2.76	-1.13 ±0.98	0.91	0.25	-0.74	1.35	9	6.29
Unionized fractions	0.30 ±0.30	0.53 ±0.39	0.54 ±0.28	-2.01 ±0.26	1.54 ±0.18	1.06 ±0.21	0.85	0.41	0.73	0.55	52	72.07
All compounds at pH 7	0.23 ±0.34	0.27 ±0.42	-0.37 ±0.21	-1.45 ±0.25	1.77 ±0.20	1.08 ±0.23	0.84	0.44	0.81	0.49	61	59.12
Humic acids in literature	1.35 ±0.66	0.76 ±0.73	-1.20 ±0.36	-2.12 ±0.48	-0.02 ±0.67	1.64 ±0.40	0.82	0.59	0.77	0.70	33	25.30
Soils & Sediments in literature	1.53 ±0.64	1.14 ±0.85	-0.89 ±0.32	-2.38 ±0.44	-0.29 ±0.53	1.36 ±0.35	0.86	0.49	0.80	0.61	35	36.16

See Table 4.1.6 explanation, A parameter was calculated by Platts et al. (1999).

Table 4.1.9 shows the application of the LSER prediction approach to investigate the interaction mechanisms of sorption of hydrophobic organic compound onto different types of HAs and soils. The sorbent specific descriptors of sorption have been found in literature from different proposed LSER models by different authors. The values of sorbent descriptors are shown in Table 4.1.9. It was found that the predominant sorption mechanisms to organic matter are hydrophobic partitioning, van-der- Waals forces and hydrogen bond formation.

Table 4.1.8 Results of linear regression of log K_{oc} modelling for different chemical classes.

Substances	Model Equation	r^2	rms	q_{cv}^2	rms_{cv}	N	F-Test
Anilines & R-benzene (R= H, CH$_3$, 3(CH$_3$), CHO, and F)	log K_{oc} = (-3.82±0.85) E_{est} + (4.48±0.52)V	0.99	0.07	0.52	0.57	8	33.60
Organochlorines & biphenyls & chlorobenzenes	log K_{oc} = (0.63±0.08) log K_{ow} + (1.17±0.46)	0.84	0.36	0.82	0.42	15	68.42
Organochlorines & biphenyls & chlorobenzenes	log K_{oc} = (6.58±2.19) E_{est}−(3.85±1.66)S_{est} +(1.39±0.38)	0.92	0.25	0.78	0.46	15	21.71
PAHs & substituted PAHs	log K_{oc} = (0.67±0.21) log K_{ow}	0.50	0.57	0.44	0.69	12	10.21
PAHs & substituted PAHs	log K_{oc} = (0.57±0.15) log K_{ow}+(1.25±0.38)S_{est}	0.78	0.39	0.67	0.53	12	15.59
Phenols	log K_{oc} = (0.42±0.13) log K_{ow} +(1.73±0.52)	0.48	0.57	0.41	0.32	14	10.96
Phenols	log K_{oc} = (0.34±0.11) log K_{ow}−(1.70±0.70)A_{est} +(3.26±0.76)	0.66	0.46	0.49	0.64	14	10.71
Phenols	log K_{oc} = (-1.47±0.54) A_{est} +(1.52±0.34)V +(1.91±0.72)	0.88	0.28	0.56	0.59	14	11.43
Heterocyclic compounds	log K_{oc} = (0.34±0.14) log K_{ow} +(1.97±0.39)	0.37	0.41	-0.19	0.64	12	5.88

N is no. of points, r^2 is squared correlation coefficient, q_{cv}^2 is predictive squared correlation coefficient using leave-one-out cross validation, calculated as shown in Appendix 9.21 (Schüürmann et al. 2008), rms is root-mean-square error of correlation, rms_{cv} is root-mean-square error of prediction using leave-one-out cross validation, F-test was calculated at α = 0.05.

Figure 4.1.6 shows that LSER exp. input descriptors have provided the best performance for neutral compounds. The K_{ow} approach and LSER exp. input descriptors failed to provide a reliable prediction for both neutral acids and bases, in contrast to LSER calc. input descriptors. Figure 4.1.7 shows that the linear fit of experimental log K_{oc} has been improved after exclusion of nine ionized compounds of f_u < 0.995 and remodelling for only fifty-two unionized fractions. The LSER approach has been found a more reliable predictor in the modelling of experimental log K_{oc} for unionized compounds than the K_{ow} classical approach.

Table 4.1.9 Comparison with the sorbent specific descriptors in literature.

Sorbent	Ref.	e	l	v	b	a	s	c
Aldrich HA	Current study-exp. input	0.48 ±0.22		1.68 ±0.24	-1.66 ±0.27			1.15 ±0.27
HAs literature	Current study-exp. input	0.83 ±0.36		1.03 ±0.48	-2.25 ±0.47	-1.43 ±0.51		1.69 ±0.43
Soil literature	Current study-exp. input	0.80 ±0.32		1.19 ±0.38	-2.40 ±0.37	-0.98 ±0.44		1.31 ±0.36
Aldrich HA	Neale et al. (2012)	0.59 ±0.08		3.94 ±0.16	-3.40 ±0.16	0.63 ±0.19	-0.52 ±0.12	-0.85 ±0.16
Pahokee Peat	Bronner et al. (2010)	0.81 ±0.08		2.99 ±0.11	-3.44 ±0.18	-0.21 ±0.14	-0.61 ±0.11	-0.29 ±0.12
Pahokee Peat	Endo et al. (2009a)	0.31 ±0.20		3.71 ±0.22	-3.94 ±0.30	-0.10 ±0.25	1.27 ±0.29	-1.04 ±0.24
Leonardite HA	Niederer et al. (2007)		0.81 ±0.07	-0.08 ±0.27	1.88 ±0.15	3.62 ±0.13	1.14 ±0.17	-0.65 ±0.15
Amherst HA	Niederer et al. (2007)		0.55 ±0.11	0.39 ±0.41	1.81 ±0.24	3.39 ±0.25	1.26 ±0.26	-0.20 ±0.24
Waskish Peat HA	Niederer et al. (2007)		0.60 ±0.10	0.51 ±0.40	1.78 ±0.23	3.68 ±0.25	0.95 ±0.26	-0.82 ±0.25
Nordic Lake HA	Niederer et al. (2007)		0.64 ±0.16	0.08 ±0.57	1.34 ±0.33	3.27 ±0.28	0.71 ±0.38	-0.35 ±0.29
Suwannee River HA	Niederer et al. (2007)		0.54 ±0.13	0.35 ±0.48	1.41 ±0.26	3.40 ±0.28	0.85 ±0.32	-0.35 ±0.29
Aldrich HA	Niederer et al. (2007)		0.94 ±0.12	-1.09 ±0.48	2.67 ±0.27	3.13 ±0.29	0.48 ±0.33	-0.38 ±0.31
Leonardite HA	Niederer et al. (2006)		0.23	2.52	-3.08	-0.49	-0.93	0.27
Soil literature	Nguyen et al. (2006)	1.10 ±0.10		2.28 ±0.14	-1.98 ±0.14	0.15 ±0.15	-0.72 ±0.14	0.14 ±0.10
Soil literature	Poole et al. (1999)	0.74 ±0.04		2.09 ±0.10	-2.27 ±0.11	-0.31 ±0.09		0.21 ±0.09

The coefficients e (solvent dispersion interaction), s (the ability of solvent phase to undergo dipole-dipole and dipole-induced dipole interactions with a solute), a (the complementary solvent hydrogen bond basicity), b (the solvent phase hydrogen bond acidity), v (endoergic cavity effect + exogeric solute-solvent effect), l (the logarithm of hexadecane/air partition constant), and the constant c (solvent specific free energy contribution), are calculated from solutes descriptors via multiparameter linear regression analysis of the dataset.

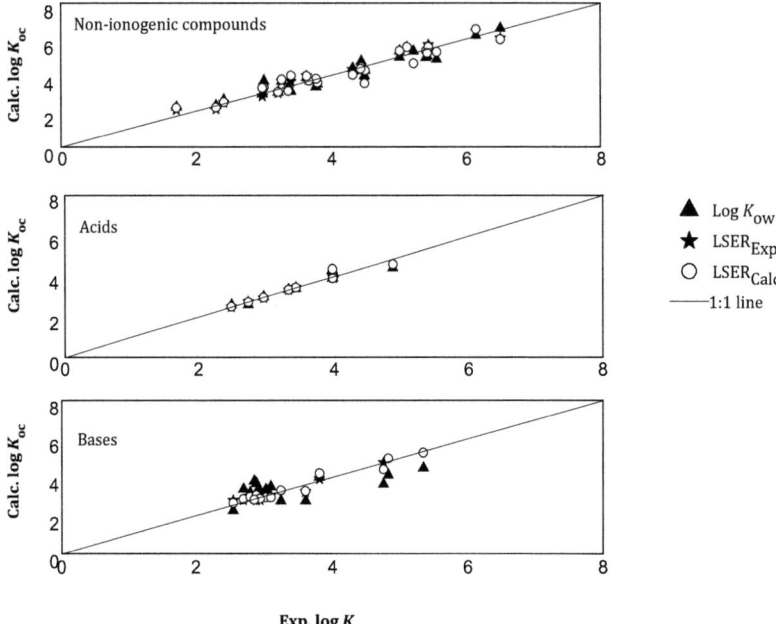

Figure 4.1.6 Results of linear regression of calc. log K_{oc} versus exp. log K_{oc} using different models at 25 ± 2 °C and pH 7 ± 0.20 for non-ionogenic, acidic and basic compounds.

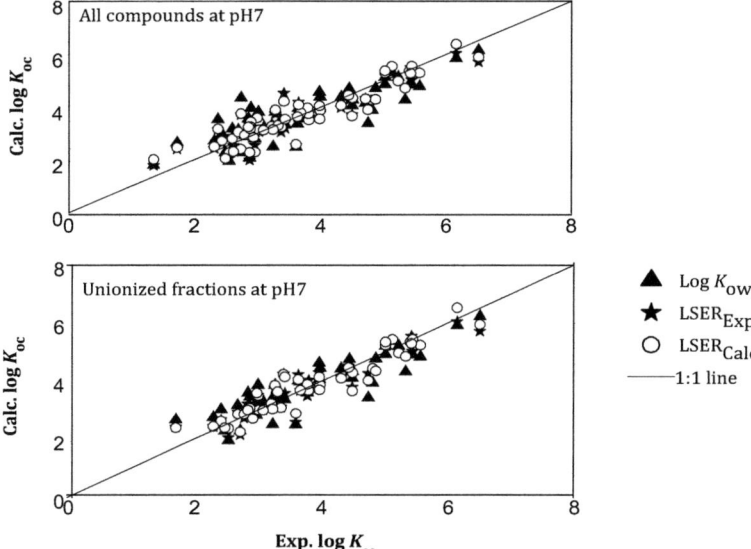

Figure 4.1.7 Results of linear regression of calc. log K_{oc} versus exp. log K_{oc} using different models at 25 ± 2 °C for all compounds and unionized fractions at pH7.

4.1.6 Conclusions

The obtained data of K_d and K_{oc} for 61 compounds in our study will help to assess their biodegradation, bioaccumulation and environmental risks. The experimental log K_{oc} values for a number of new compounds (1-aminopyrene, 2,3-dimethylnaphthalene, 2,4`-DDD, 2,4`-DDT, 2,4-di-t-butylphenol, 2-fluorobiphenyl, 2-methylanthracene, 2-nitrofluorene, 3-phenylphenol, 4,4`-isopropylidenediphenol, 4-n-hexylphenol, 4-n-octylphenol, 6-nitroquinoline and sebuthylazine) are not found in literature but were determined in our study. Log K_{oc}–log K_{ow} correlation was insignificant for bases, heterocyclic compounds and phenols due to protonation and deprotonation depending on their pK_a values but recommended for organochlorines, biphenyls and chlorobenzenes. By accompanying log K_{ow} with S_{est}, the polarizability property of the solute, a slight improvement of the prediction of sorption for organochlorines and PAHs was observed due to π-π interactions (interactions), in which PAHs and organochlorines act as electron donors and the aromatic moieties of HSs act as electron acceptors. The prediction of log K_{oc} of phenols was improved by applying of LSER calc. input only. An improvement of the prediction performance by both of K_{ow} and LSER prediction models for K_{oc} was achieved by excluding ionized compounds from the dataset which have unionized fractions f_u < 0.995. LSER approach can be recommended to neutral species of phenols, neutral species of anilines, R-benzenes and unionized fractions of hydrophobic compounds. Both prediction methods log K_{ow} and LSER do not fit for ionisable compounds. Measurement limitations of the batch technique and SPME-GC-MSD have been found with polar compounds due to speciation and difficult uptake to the fibre. Hydrophobic partitioning, hydrogen bonding and electron donor-acceptor interactions (EDA interactions) have been found to be the dominant sorption interactions between organic solutes and HA.

4.2 pH-Dependence of sorption

4.2.1 Introduction

The major objectives of part 4.2 are: First, to investigate the effect of pH variation on the experimental values of log K_{oc} according to deprotonation and protonation of the solute in the selected range of pH and pK_a and to discuss the sorption mechanisms. Second, to create a log K_{oc}-pH dependent prediction model according to the experimental data that can give good predictions that will fit for each category of selected organic chemicals (neutral, acids and bases). Third, to compare with other different prediction methods in literature. Forth, to show the regression performance of LSER approaches of sorption at different pH values, where it is not completely investigated in literature. In this study, thirty-seven substances as first subset were selected from Scheme 3.1. Different types of organic solutes were chosen to be under investigation (PAHs, organochlorine, anilines, phenols, and triazines) by using direct immersion-mode-nd-SPME-GC-MSD.

4.2.2 Comparison with mean log K_{oc} in literature

Log K_{oc} data of substances have been measured by direct immersion mode as shown in Appendix 9.10, they were calculated from equation (3.8) in chapter 3. Table 4.2.1 shows experimental values of log K_{oc} at different pH solutions with their maximum error. The exp. log K_{oc} values in the current study have been measured at 25±2 °C and at different pH and only the data at pH 7 compared with three types of mean values of log K_{oc} in selected literature at pH 7 as shown in Appendix 9.16. That data is related to different sorbents, firstly to different HAs, secondly to various soils and sediments, and thirdly to the non-specific data which was collected from references without indication of their original sorbent types (different OM). For six compounds in this subset, no data is available in literature (G2, K3, K4, L1, O2 and P1) as shown in Appendix 9.16.

As mentioned before in chapter 4.1, the wide range of log K_{oc} in literature has been expected due to different solute-sorbent (solvent) interactions including different contribution of intermolecular interactions. The sorption strength onto HAs in literature is only slightly higher unit than that of soils and sediments as shown in chapter 4.1. This very small difference in sorption may be attributed to similar types of molecular interactions or similar organic carbon content. It indicates that the sorption is expressed as a function of the organic carbon content of the sorbent (Doucette 2003) without neglecting the differences in organic matter properties and the magnitude of non-organic matter contribution to the sorption.

Figure 4.2.1 shows that evidence: at low K_{ow} and K_{oc} the sorption capacity of soils and sediments is higher than those of HAs while at higher K_{ow} and K_{oc} for high hydrophobic compounds, the sorption capacity of HAs is higher than those of soils and sediments. This may be due to the difference in sorption intermolecular interactions between low hydrophobic and high hydrophobic compounds with OM and may be also with mineral surface. Low hydrophobic compounds can interact with both of MO and the mineral surface of soil and sediments predominantly through hydrogen bonding and charge transfer while high hydrophobic compounds can interact with OM (in the current study is HA) via van-der-Waals and hydrophobic bonding.

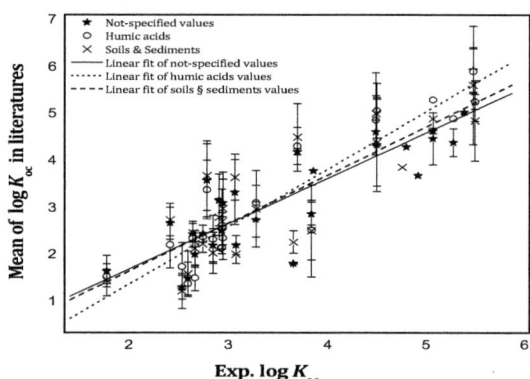

Figure 4.2.1 Comparison between mean values of exp. log K_{oc} in literature and exp. log K_{oc} in current study. Not-specified values are not relevant to specific sorbent.

Table 4.2.1 The experimental values of sorption coefficients for selected subset at 25±2 °C and at different pH values (f_{oc} of used Aldrich HA is 0.395 g/g).

No.	Substance	CAS No.	Log K_{ow}	pK_a	pH	$K_{d(mean)}$ [L/kg]	Log K_d ± Err	$K_{oc(mean)}$ [L/kg$_{oc}$]	Log K_{oc} ±Err
				Neutral					
1	Benzene	71-43-2	2.13	n.a.	4	27	1.43±0.05	68	1.83±0.05
					7	21	1.33±0.03	54	1.73±0.03
					10	10	1.02±0.02	26	1.42±0.02
2	1,3,5-Trichlorobenzene	108-70-3	4.19	n.a.	4	950	2.98±0.11	2399	3.38±0.11
					7	415	2.62±0.11	1047	3.02±0.11
					10	387	2.59±0.12	977	2.99±0.12
3	Hexachlorobenzene	118-74-1	5.73	n.a.	4	24417	4.39±0.05	61659	4.79±0.05
					7	11161	4.05±0.07	28184	4.45±0.07
					10	6725	3.83±0.23	16982	4.23±0.23
4	Naphthalene	91-20-3	3.30	n.a.	4	387	2.59±0.08	977	2.99±0.08
					7	672	2.83±0.21	1698	3.23±0.21
					10	886	2.95±0.48	2239	3.35±0.48
5	2,3-Dimethylnaphthalene	581-40-8	4.40	n.a.	4	1438	3.16±0.04	3630	3.56±0.04
					7	1895	3.28±0.18	4786	3.68±0.18
					10	1041	3.02±0.13	2630	3.42±0.13
6	Anthracene	120-12-7	4.45	n.a.	4	2030	3.31±0.04	5129	3.71±0.04
					7	1769	3.25±0.05	4467	3.65±0.05
					10	1729	3.24±0.11	4365	3.64±0.11
7	Fluoranthene	206-44-0	5.16	n.a.	4	13418	4.13±0.09	33884	4.53±0.09
					7	10907	4.04±0.06	27542	4.44±0.06
					10	10658	4.03±0.06	26915	4.43±0.06

Table 4.2.1 Continued.

No.	Substance	CAS No.	Log K_{ow}	pK_a	pH	$K_{d(mean)}$ [L/kg]	Log K_d ± Err	$K_{oc(mean)}$ [L/kg$_{oc}$]	Log K_{oc} ±Err
8	Benzo(a)pyrene	50-32-8	6.13	n.a.	4	337051	5.53±0.03	851138	5.93±0.03
					7	106585	5.03±0.07	269153	5.43±0.07
					10	53419	4.73±0.08	134896	5.13±0.08
9	Aldrin	309-00-2	6.50	n.a.	4	53419	4.73±0.09	134896	5.13±0.09
					7	67250	4.83±0.15	169824	5.23±0.15
					10	90718	4.96±0.21	229087	5.36±0.21
10	Lindane	58-89-9	3.72	n.a.	4	808	2.91±0.02	2042	3.31±0.02
					7	2442	3.39±0.09	6166	3.79±0.09
					10	2386	3.38±0.15	6025	3.78±0.15
11	4,4`-DDD	72-54-8	6.02	n.a.	4	54663	4.74±0.10	138038	5.14±0.10
					7	41467	4.62±0.05	104715	5.02±0.05
					10	80853	4.91±0.07	204174	5.31±0.07
12	4,4`-DDE	72-55-9	6.51	n.a.	4	54663	4.74±0.05	138038	5.14±0.05
					7	41469	4.62±0.07	104720	5.02±0.07
					10	80853	4.91±0.14	204174	5.31±0.14
13	4,4`-DDT	50-29-3	6.91	n.a.	4	111608	5.05±0.12	281838	5.45±0.12
					7	111608	5.05±0.11	281838	5.45±0.11
					10	75456	4.88±0.09	190546	5.28±0.09
			Acids						
14	Phenol	108-95-2	1.46	9.99	4	49	1.69±0.19	123	2.09±0.19
					7	122	2.09±0.20	309	2.49±0.20
					10	9	0.96±0.07	23	1.37±0.07

Table 4.2.1 Continued.

No.	Substance	CAS No.	Log K_{ow}	pK_a	pH	$K_{d(mean)}$ [L/kg]	Log K_d ± Err	$K_{oc(mean)}$ [L/kg$_{oc}$]	Log K_{oc} ±Err
15	4-Nitrophenol	100-02-7	1.91	7.15	4	66	1.82±0.14	166	2.22±0.14
					7	165	2.22±0.13	417	2.62±0.13
					10	n.d.	n.d.	n.d.	n.d.
16	2,4-Dichlorophenol	120-83-2	3.06	7.89	4	218	2.34±0.17	549	2.74±0.17
					7	154	2.19±0.03	390	2.60±0.03
					10	154	2.19±0.10	389	2.59±0.10
17	2,4,6-Trichlorophenol	88-06-2	3.69	6.23	4	300	2.48±0.12	758	2.88±0.12
					7	93	1.97±0.34	234	2.37±0.34
					10	n.d.	n.d.	n.d.	n.d.
18	2,3,4,6-Tetrachlorophenol	58-90-2	4.45	5.22	4	827	2.92±0.02	2089	3.32±0.02
					7	314	2.50±0.30	794	2.90±0.30
					10	n.d.	n.d.	n.d.	n.d.
19	Pentachlorophenol	87-86-5	5.12	4.70	4	1041	3.02±0.16	2630	3.42±0.16
					7	218	2.34±0.43	549	2.74±0.43
					10	n.d.	n.d.	n.d.	n.d.
20	2,4-Di-t-butylphenol	96-76-4	5.19	11.70*	4	2498	3.40±0.06	6309	3.80±0.06
					7	3870	3.59±0.12	9772	3.99±0.12
					10	4052	3.61±0.06	10232	4.01±0.06
21	3-Phenylphenol	580-51-8	3.23	9.64	4	534	2.73±0.08	1349	3.13±0.08
					7	847	2.93±0.10	2138	3.33±0.10
					10	194	2.29±0.23	490	2.69±0.23
22	4,4`-Isopropylidenediphenol	80-05-7	3.32	10.29** 10.93**	4	1018	3.01±0.01	2570	3.41±0.01
					7	1091	3.04±0.22	2754	3.44±0.22
					10	1066	3.03±0.31	2691	3.43±0.31

Table 4.2.1 Continued.

No.	Substance	CAS No.	Log K_{ow}	pK_a	pH	$K_{d(mean)}$ [L/kg]	Log K_d ± Err	$K_{oc(mean)}$ [L/kg$_{oc}$]	Log K_{oc} ±Err
23	4-n-Nonylphenol	104-40-5	5.76	10.15**	4	32938	4.52±0.02	83177	4.92±0.02
					7	29356	4.47±0.05	74131	4.87±0.05
					10	24417	4.39±0.02	61659	4.79±0.02
24	Triclosan	3380-34-5	4.76	7.80**	4	12814	4.11±0.04	32359	4.51±0.04
					7	20309	4.31±0.17	51286	4.71±0.17
					10	2176	3.34±0.08	5495	3.74±0.08
Bases
25	Aniline	62-53-3	0.90	4.60	4	280	2.45±0.15	708	2.85±0.15
					7	140	2.15±0.39	355	2.55±0.39
					10	250	2.40±0.15	631	2.80±0.15
26	4-Chloroaniline	106-47-8	1.83	3.98	4	83	1.92±0.36	209	2.32±0.36
					7	1613	3.21±0.06	4074	3.61±0.06
					10	455	2.66±0.09	1148	3.06±0.09
27	2,3,5,6-Tetrachloraniline	3481-20-7	4.10	-1.40**	4	4243	3.63±0.01	10715	4.03±0.01
					7	2557	3.41±0.05	6456	3.81±0.05
					10	3449	3.54±0.01	8710	3.94±0.01
28	Ametryn	834-12-8	2.98	4.10	4	n.d.	n.d.	n.d.	n.d.
					7	198	2.30±0.23	501	2.70±0.23
					10	154	2.19±0.33	390	2.60±0.33
29	Atrazine	1912-24-9	2.61	1.70	4	194	2.29±0.31	490	2.69±0.31
					7	250	2.40±0.11	631	2.80±0.11
					10	322	2.51±0.07	813	2.91±0.07

Table 4.2.1 Continued.

No.	Substance	CAS No.	Log K_{ow}	pK_a	pH	$K_{d(mean)}$ [L/kg]	Log K_d ± Err	$K_{oc(mean)}$ [L/kg$_{oc}$]	Log K_{oc} ±Err
30	Prometryn	7287-19-6	3.51	4.05	4	194	2.29±0.09	490	2.69±0.09
					7	307	2.49±0.05	776	2.89±0.05
					10	369	2.57±0.14	933	2.97±0.14
31	Propazine	139-40-2	2.93	1.70	4	250	2.40±0.10	631	2.80±0.10
					7	424	2.63±0.03	1071	3.03±0.03
					10	642	2.81±0.10	1621	3.21±0.10
32	Sebuthylazine	7286-69-3	2.61*	2.50**	4	244	2.39±0.03	616	2.79±0.03
					7	307	2.49±0.28	776	2.89±0.28
					10	274	2.44±0.08	692	2.84±0.08
33	Terbutryn	886-50-0	3.74	4.03**	4	244	2.39±0.07	616	2.79±0.07
					7	287	2.46±0.03	724	2.86±0.03
					10	487	2.69±0.05	1230	3.09±0.05
34	2-Methylpyridine	109-06-8	1.11	6.00	4	n.d.	n.d.	n.d.	n.d.
					7	300	2.48±0.13	758	2.88±0.13
					10	154	2.19±0.10	390	2.60±0.10
35	2-Aminoanthracene	613-13-8	3.43#	4.32**	4	83	1.92±0.04	209	4.73±0.04
					7	22269	4.35±0.05	56234	4.75±0.05
					10	18954	4.28±0.02	47863	4.68±0.02
36	1-Aminopyrene	1606-67-3	4.31	4.32**	4	72060	4.86±0.17	181970	5.26±0.17
					7	26163	4.42±0.19	66069	4.82±0.19
					10	34490	4.54±0.16	87096	4.94±0.16
37	6-Aminochrysene	2642-98-0	4.99	4.32**	4	28034	4.45±0.12	70794	4.85±0.12
					7	86635	4.94±0.02	218776	5.34±0.02
					10	30040	4.48±0.04	75858	4.88±0.04

Err is maximum logarithmic error of exp. K_{oc} and K_d, it was calculated as shown in Appendix 9.20.(#) is estimated log K_{ow} by Epi suite ™ v.4 (KOWWIN software). (**) is estimated pK_a by ACD/pK_a v.12 software. (*) is experimental log K_{ow} from Environmental science-Interactive physprop database of Syracuse Research Corporation (SRC). (n.d.) is not detected. (n.a.) is not available. The exp. log K_{oc} values of underlined substances was calculated from peak area (Method II in chapter 3) due to the very small peaks of their calibration curves. f_{oc} of used Aldrich HA is 0.395 g/g (see chapter 3).

4.2.3 Comparison with calculated log K_{oc} from different prediction methods

The exp. log K_{oc} values in the current study which were measured at 25±2 °C have been compared with the calculated log K_{oc} values using different prediction methods in literature as shown in Appendix 9.17. Overall, the statistics summarized in Table 4.2.2 have shown the linear regression of calc. log K_{oc} versus exp. log K_{oc} at pH 7. The fragment constant model by Tao et al. (1999) is statistically robust and has the best prediction performance compared to the results achieved with the literature methods. It was found a difference of 0.09 between r^2 and q^2 indicating the absence of systematic errors and a diverse training set in the Tao model covers most of the calibration dataset of current study including different chemical classes as shown in the training set and chemical domain check of each model in Appendix 9.22. That model calculates log K_{oc} from contributions by individual atoms, bonds, functional groups that have been fitted to a dataset. In addition, the Tao model has the lowest *rms*.

The second best fitting performance was the calc. log K_{oc} of KOCWIN software using molecular topology and fragment contribution method (Meylan et al. 1992). This model has shown the lowest difference between r^2 and q^2 of 0.04 indicating the high quality of its original training set and the absence of systematic error. The low regression coefficient r^2 of calc. log K_{oc} based on log K_{ow} by KOCWIN has been attributed to the model not fitting polar compounds whose K_{oc} values can be affected by protonation or deprotonation depending on pH variation. The third best fitting performance was that of the 2D molecular structure model (Schüürmann et al. 2006) with a low difference between r^2 and q^2 of 0.05 indicating the absence of systematic errors of the model and may be due to a wide range training set regarding the chemical domain in the Schüürmann model.

LSER approach by Poole et al. (1999) for both calc. and exp. inputs have shown the fourth and fifth best fitting performance with a substantial difference between r^2 and q^2 of 0.24 and 0.18. Franco and Trapp model (2008) was applied for only unionized fractions at pH7 with consideration $pK_a \simeq 50$ for non-polar compounds. It was found 0.25 difference between r^2 and q^2 with high error *rms*= 0.75. The calibration dataset of Franco and Trapp model (2008) did not include chlorobenzenes, PAHs derivatives and PCBs. The similar Nguyen model (2005) for both calc. and exp. input has shown a large difference between r^2 and q^2, with 0.25 and 0.30 respectively. Experimental Abraham descriptors are not available for all chemical compounds, which can be a problem in such prediction modelling. Calc. log K_{oc} of KOCWIN software based on log K_{ow} method has shown a very low fitting performance with 0.16 difference between r^2 and q^2

associated with high root-mean-square error.

The molecular connectivity indices model (Sabljic' et al. 1995) has shown the low statistical performance with a high difference between r^2 and q^2 of 0.18 associated with the high rms = 0.77 due to presence of a systematic error. The last two prediction models show a not well fitted regression, as these prediction models neglect the effect of pH variation for polar compounds. As the log of their K_{ow} data varies due to protonation and deprotonation, the neutral species amount will decrease. This makes it a poor predictor in such cases. Recent LSER models such as Endo et al. (2009a), Bronner al. (2010a) and Neale et al. (2012) have been used for both experimental and calculated solute descriptors. It was found similar behaviour for these methods, a large gap between r^2 and q^2 with high value of rms and negative q^2 that indicates the poor predictability due to their calibration dataset does not cover all used solutes and presence of systematic errors.

4.2.4 Refitting of the common K_{ow}-approach for sorption to humic acid

The sorption process onto soil can be considered as a partitioning process between water and a hydrophobic soil phase. One of the prediction models for soil sorption, the classical approach, has used octanol as surrogate for the organic carbon in soils, sediments and OM (Gawlik et al. 1997 and Schüürmann et al. 2007). The classical model for prediction of log K_{oc} from the log K_{ow} is very important for neutral species because this model is pH-independent. Figure 4.2.2 shows that log K_{ow}-log K_{oc} correlations for 37 substances of different chemical structures with a wide range of hydrophobicity log K_{ow} (0.90 – 6.91). There is no high fitting of log K_{oc} for the dataset that has been found for total neutral compounds, acids and bases and included different chemical structures such as phenols, heterocyclic, anilines, organochlorines and PAHs. It shows also that acids and bases do not fit well with the K_{ow} approach to predict K_{oc}, and some points are outside the 95% confidence limit of the linear fit which may be attributed to their hydrohobicities being an insufficient descriptor for soil adsorption process, and that the effect of pH-pK_a correlation is the reason for speciation.

The log K_{ow}-log K_{oc} correlations (Karickhoff et al. 1981, Sabljić et al. 1995, Gawlik et al. 1997 and Doucette 2003) at pH 4, 7, and 10 ± 0.2 at 25±2°C have been performed and their linear regression results with full statistics have been calculated and illustrated in Table 4.2.3. The experimental dataset of log K_{ow} has been used in modelling and in cases where no experimental data is available, the calculated log K_{ow} has been determined using KOWWIN software.

Table 4.2.2 Results of linear regression of calc. log K_{oc} versus exp. log K_{oc} using different methods in literature.

Model	N	r^2	q^2	rms	F-Test	bias	me	mne	mpe
2D Molecular structure[a]	37	0.70	0.65	0.61	92.74	-0.20	0.47	-1.46	1.36
KOCWIN[b] Molecular topology (top indices +fragments)	37	0.78	0.74	0.53	158.10	-0.05	0.39	-1.56	1.14
KOCWIN[b] (log K_{ow})	37	0.65	0.49	0.74	90.73	-0.25	0.55	-1.77	1.33
Molecular connectivity indices[c]	37	0.62	0.44	0.77	78.21	-0.33	0.61	-1.73	0.94
Fragment constant model[d]	37	0.84	0.75	0.52	172.75	-0.25	0.41	-1.21	0.91
LSER (Nguyen)[e] exp.	29#	0.67	0.37	0.73	80.45	-0.04	0.53	-1.97	1.29
LSER (Nguyen)[e] calc.	37	0.73	0.48	0.74	118.64	-0.22	0.58	-1.65	1.59
LSER (Poole)[f] exp.	29#	0.74	0.56	0.62	96.77	-0.18	0.45	-1.61	0.88
LSER (Poole)[f] calc.	37	0.77	0.53	0.71	121.79	-0.33	0.59	-1.66	1.10
Franco & Trapp model (2008)[g]	30*	0.72	0.47	0.75	49.15	-0.50	0.59	-1.79	0.53
LSER (Endo 2009 – low Peat)[h] exp.	37	0.70	-1.34	1.59	61.46	1.20	1.42	-1.63	2.93
LSER (Endo 2009 – low Peat)[i] calc.	37	0.78	-1.84	1.75	65.55	1.22	1.45	-1.39	3.58
LSER (Endo 2009 – high Peat)[j] exp.	37	0.72	0.02	1.03	98.94	0.28	0.79	-2.26	2.47
LSER (Endo 2009 – high Peat)[k] calc.	37	0.71	-0.29	1.18	91.54	-0.05	0.96	-2.00	2.14
LSER (Bronner 2010a)[l] exp.	37	0.65	-0.68	1.34	71.77	-0.71	1.08	-2.44	1.43
LSER (Bronner 2010a)[m] calc.	37	0.65	-0.68	1.34	71.77	-0.71	1.08	-2.44	1.43
LSER (Neale 2012)[n] exp.	37	0.68	-0.68	1.35	75.53	0.68	1.07	-2.17	3.20
LSER (Neale 2012)[o] calc.	37	0.66	-0.44	1.24	83.11	0.26	1.03	-1.91	2.83

a) Schüürmann et al. 2006. b) KOCWIN software. c) Sabljić et al. 1995. d) Tao et al. 1999. e) Nguyen et al. 2005. f) Poole et al. 1999. g) Franco et al. 2008. h-k) Endo et al. 2009a. l-m) Bronner et al. 2010a. n-o) Neale et al. 2012. #Experimental Abraham descriptors were not available for all solutes (see Appendix 9.13, 9.14 and 9.17). * The model was used only for 30 unionized fractions in pH 7± 0.20. r^2 is squared correlation coefficient, q^2 is predictive squared correlation coefficient was calculated as shown in Appendix 9.21 (Schüürmann et al. 2008), rms is root-mean-square error of correlation, bias is systematic error, me is mean error, mne is maximum negative error, mpe is maximum positive error and F-test was calculated at α = 0.05.

Log K_{oc}-log K_{ow} correlation for neutral, acids and bases have been shown in Figure 4.2.3 and Table 4.2.3. The results show an agreement with Sabljić et al. (1995) models, indicating a weak contribution of hydrophobicity parameters in the sorption process, which can be attributed to the fact that the hydroxyl group of phenols (acid model compounds) and electron pair of the nitrogen atoms of heterocyclic and amino group (base model compounds) can be used as hydrogen bond acceptor, so the sorption can occur via non-dispersive interactions. This result agrees with literature.

Feng et al. (1996) have shown that the difficulty to apply the K_{ow}-K_{oc} correlation for polar compounds can be attributed to the fact that the organic carbon phase is mainly responsible for adsorption to soil where there are more cohesive and stronger hydrogen bonds donor solvents than n-octanol. The sorption process of polar or ionizable compounds depends on non-hydrophobic and non-dispersive interactions.

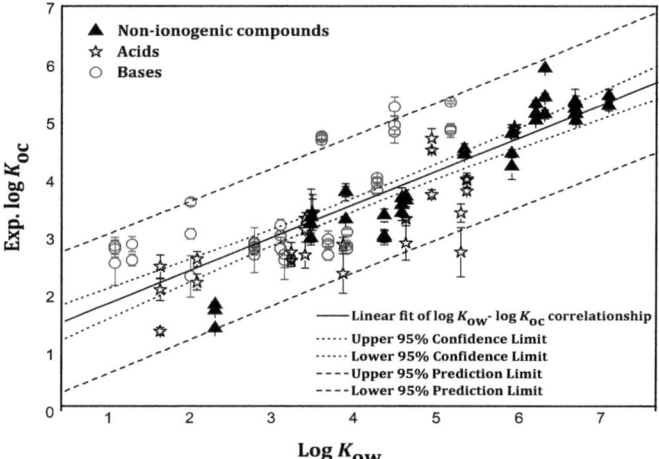

Figure 4.2.2 Exp. log K_{oc} versus log K_{ow} for non-ionogenic compound, acids and bases at different pH values.

As mentioned before in chapter 4.1, the applicability to use K_{ow} to predict K_{oc} is not reliable for polar compounds (von Oepen et al. 1991 and Pussemier et al. 1989) due to the underestimation of log K_{oc} for acids and the overestimation of log K_{oc} for bases.

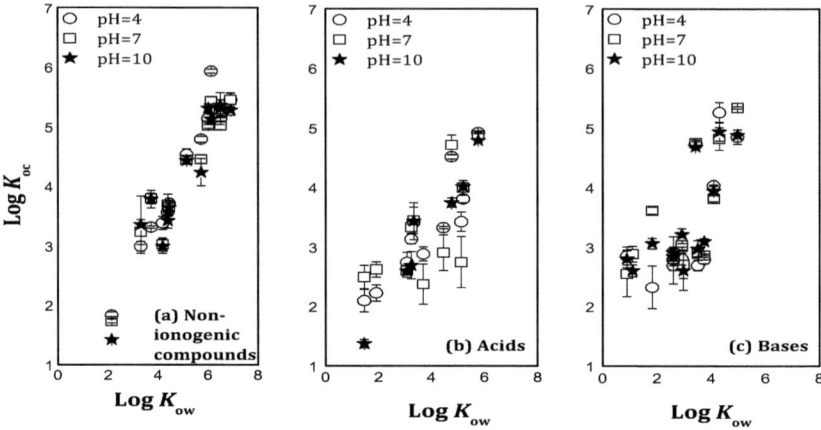

Figure 4.2.3 Log K_{oc} versus exp. log K_{ow} for non-ionogenic compounds, acids and bases at pH 4, 7 and 10.

Table 4.2.3 shows log-log correlations between K_{oc} and K_{ow} for each pH value for non-ionogenic compounds, acidic and basic compounds. The fact that the common log K_{ow} model fitted only with non-ionogenic compounds was proven by highly significant statistics, correlation and predictive coefficients (r^2) and (q_{cv}^2). The hydrophobicity descriptor (K_{ow}) is sufficient for such compounds. That model fitted well and showed a robust statistics. Highly significant correlation and predictive coefficients were calculated by using the ChemProp program. Log K_{oc} Benzo(a)pyrene at pH4 was outlier because of enhancement of sorption onto HSs due to highly π-π interactions with aromatic moieties of HA especially with high hydrophobic compounds of benzoic structure. Nevertheless, there were low correlation and prediction coefficients for acids at pH 4 and 7 and for bases at pH 4, 7 and 10. In cases of dissociation or protonation, it can occur that the model will fail to predict the right values of log K_{oc} because the sorption mechanism does not depend only on hydrophobicity parameters, but the molecular size and polarizability also play important roles. It has been found that the correlation coefficient of multiple linear regressions and the prediction performance will be improved and highly significant after exclusion of the ionized compounds whose (the unionized fractions f_u and the ionized fractions f_i) $f_u < 0.995$.

This presents a case that such models cannot be used for prediction of polar substances which can be protonated or deprotonated. This is the reason why we are interested in setting up a preliminary model to predict the sorption capacity of polar and non-polar substances. This is important for the investigation of the transport, mobility and environmental fate of different types of organic solutes.

The acids and bases behaviour in water varies according to their pK_a and pH of solution, as described in the following equations (4.2.1) and (4.2.2):

Equation (4.2.1) describes the deprotonation of acids at pH > pK_a

$$AH \rightleftharpoons A^- + H^+ \qquad (4.2.1)$$

Thus, some acids have no measured log K_{oc} as a result of their deprotonation at pH 10 and most of the used acids are neutral at pH 4.

While equation (4.2.2) describes the deprotonation of bases at pK_a > pH

$$BH^+ \rightleftharpoons B + H^+ \qquad (4.2.2)$$

So, the similar behaviour of deprotonation of bases can be found at pH 4 and most of the used bases are in neutral form at pH 10. At pH = pK_a, the neutral and anionic species are in equal amounts, meaning that 50 % of the compound is ionized. Figure 4.2.4 and Table 4.2.3 show that the deprotonation of most used acids and bases has occurred at pH 10 and 4 respectively. The neutral form and the ionic species are found in the same solutions and it tends to be that at pH 4 for acids, unionized fractions are more clearly observed than at pH 10 for bases.

Table 4.2.3 Results of linear regression of log K_{oc} versus log K_{ow} at different pH for non-ionogenic compounds, acids and bases.

Substances	pH	Model Equation	N	r^2	rms	q^2_{cv}	rms_{cv}	F-Test	bias	me	mne	mpe
Non-ionogenic compounds	4	log K_{oc} = (0.75±0.03) log K_{ow}	12[a]	0.94	0.28	0.93	0.35	466.13	0.00	0.14	-0.27	0.16
	7	log K_{oc} = (0.73±0.07) log K_{ow}	13	0.92	0.30	0.90	0.38	123.33	0.00	0.25	-0.57	0.54
	10	log K_{oc} = (0.77±0.08) log K_{ow}	13	0.89	0.37	0.86	0.48	92.46	0.00	0.31	-0.65	0.50
Acids	4	log K_{oc} = (0.56±0.10) log K_{ow}+ (1.19±0.39)	11	0.78	0.38	0.73	0.49	32.49	0.00	0.33	-0.67	0.62
	7	log K_{oc} = (0.43±0.16) log K_{ow}+ (1.65±0.65)	11	0.44	0.63	0.33	0.80	7.02	0.00	0.54	-1.03	1.10
	10	log K_{oc} = (0.73 ±0.08) log K_{ow}	7[b]	0.94	0.25	0.92	0.36	77.03	0.00	0.20	-0.57	0.22
Bases	4	log K_{oc} = (0.64 ±0.22) log K_{ow}	11[c]	0.49	0.72	0.32	0.97	8.63	0.00	0.61	-1.13	1.00
	7	log K_{oc} = (0.52±0.17) log K_{ow}+ (1.90±0.56)	13	0.44	0.67	0.35	0.82	8.83	0.00	0.61	-1.07	0.98
	10	log K_{oc} = (0.53±0.15) log K_{ow}+ (1.84±0.48)	13	0.54	0.57	0.47	0.69	12.77	0.00	0.49	-1.03	0.81
Unionized fractions	(4, 7, &10)	log K_{oc} = (0.59±0.04) log K_{ow}+ (1.36±0.16)	80	0.77	0.51	0.77	0.52	264.55	0.00	0.38	-1.36	1.20

a) Outlier Benzo(a)pyrene. b) J2, J4, K1, K2 were not detected at pH10. C) M4 and N2 were not detected at pH4. N is no. of points, r^2 is squared correlation coefficient, q^2_{cv} is predictive squared correlation coefficient using leave-one-out-cross validation was calculated as shown in Appendix 9.21 (Schüürmann et al. 2008), rms is root-mean-square error of correlation, rms_{cv} is root-mean-square error of prediction using leave-one-out-cross validation, bias is systematic error, me is mean error, mne is maximum negative error, mpe is maximum positive error and F-test was calculated at α = 0.05. Unionized fractions are the non-ionogenic compounds and unionized fractions of both acids and bases of $1 > f_u > 0.995$.

pH-dependence of sorption

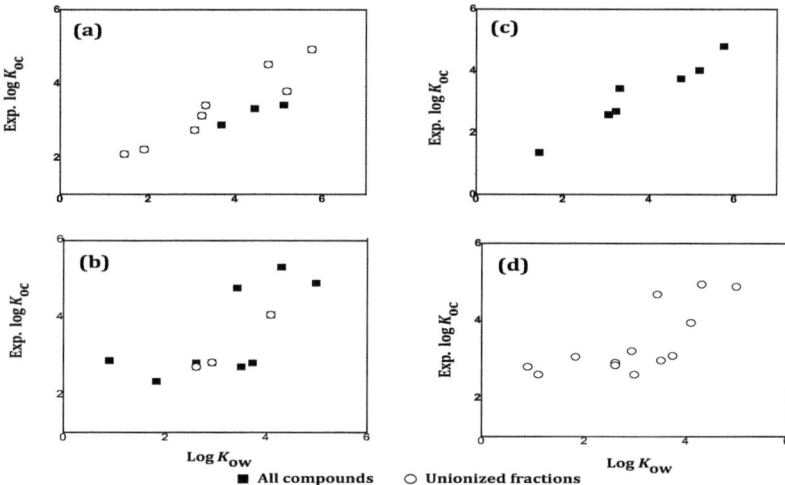

Figure 4.2.4 Log K_{oc}- log K_{ow} correlation for acids and bases for all compounds and unionized fractions. (a) acids at pH4, (b) bases at pH4, (c) acids at pH10 and (d) bases at pH10.

4.2.5 The influence of pH variation on the values of log K_{oc} (effect of solute structure, physical properties on partitioning processes according to Henderson-Hasselbach equation)

The determined partition coefficients of non-ionogenic compounds and unionized fractions of acids and bases) with their maximum errors are shown in Table 4.2.1 in the range of selected pH. Figure 4.2.5 shows the effect of pH variation on exp. log K_{oc} values of non-ionogenic solutes. The statistical T-test was applied to the experimental data of PAHs and it showed non-significant results for pH change. This may be attributed to hydrophobic bonding which took place in the hydrophobic domain of the HA and it's not being pH-dependent may be due to a high affinity of HA to interact with aromatic structure (Gauthier et al. 1986 and Xing et al. 2008). PAHs have a planar structure and the sorption of planar molecules is strongly bounded with organic matter more than that of non-planar molecules due to the relatively strong π-π interactions (electron donor - electron acceptor charge transfer) with the aromatic moieties of HA (Zhu et al. 2004 and Xing et al. 2008). Meanwhile a significant change for A4, C1 and C4 was shown and this may be attributed to the presence of high electronegative substitution as chlorine atoms have strongly negative charges and therefore can interact with HA by hydrogen bond formation or strong polarized attraction to the charge on the surface of HA (especially in acidic medium).

It will be known that the sorption of non-ionogenic substances onto humic material can be investigated through hydrophobic partitioning phenomenon, charge transfer as well as different types of van-der-Waals attraction forces such as Keesom forces, Debye forces and London dispersion forces. For D4 and F2, the T-test showed a significant change. Sorption increases under varying pH only took place between pH 7 and 10 in contrast to E1, F1 and F3 where there was only a significant change as sorption was decreasing between pH 7 and 4. This change in sorption behaviour could result from factors such as the effect of pH on deprotonation of functional group of HA due to its pK_a in the range of 3 to 6, may be the low degree of deprotonation for HA at pH 4 can occur. Other explanation can account for the behaviour of slightly increasing the sorption of neutral organic solutes such as PAHs at pH4 more than pH 7 and 10 may be due to the sorption took place in hydrophobic domain of HA which is pH- independent. By lowering pH and dissociation of functional groups of hydrophilic domain of HA occurs, thus the area of hydrophobic domain is larger than that of the hydrophilic domain due to deprotonation, then the sorption can increase. Moreover, the decrease in sorption in certain pH values may be attributed to the high the energy penalty difference of cavity formation between two phases and that drives sorption decreases. For F3, sometimes because of the high polarity of HA compared with organic solutes makes the sorption interactions less favourable (Laor et al. 2002).

Figure 4.2.6 shows that the effect of pH variation on exp. log K_{oc} values of phenols. It was observed that the log K_{oc} data decreased significantly for I4, J3, K4, and M1 due to the decrease in the concentration of the neutral form and phenolate anions formation at pH > pK_a. Moreover, in the case of J2, J4, K1, and K2 data were not detected at pH 10. The dissociation was very high and anions more than the neutral species were thus not enough for uptake by the SPME-fibre and for analysis. Moreover, we cannot deny the low degree of HA deprotonation at pH 4 which can also lower the experimental values of log K_{oc}.

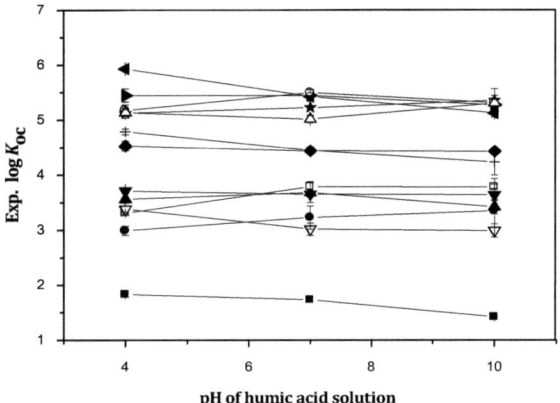

Figure 4.2.5 Exp. log K_{oc} versus pH of HA for non-ionogenic compounds.

──■── Benzene, ──●── Naphthalene, ──▲── 2,3-Dimethylnaphthalene, ──▼── Anthracene, ──◆── Fluoranthene, ──◀── Benzo(a)pyrene, ──▶── 4,4`-DDT, ──★── Aldrin, ──+── Hexachlorobenzene, ──□── Lindane, ──○── 4,4`-DDD, ──△── 4,4`-DDE, ──▽── 1,3,5-Trichlorobenzene

There was the presence of phenolate anions that sorbed very slowly due to repulsion forces between same electronic charge solution and the surface of HA in the basic medium. For unionized fractions, hydrogen bond formation and van-der-Waals attraction forces are considered the predominant mechanisms of sorption of phenol onto HA (Choudhry 1984). Sorption of anionic organic chemicals at pH values below their pK_a values can be attributed to sorption of the unionized fractions of solutes on the organic surface, thus, hydrogen bonding may be occurring between the -OH, -NH$_2$, -NH-group and -COOH or –NH- group of HA. Hydrogen bonding would be limited in acidic conditions where -COOH groups may be ionized (Stevenson 1982, Ohenbusch et al. 2000 and Rosta et al. 2003).

Insignificant changes in log K_{oc} of L1 have been shown due to its neutral form and the effect of low degree of deprotonation for HA which is so low and hence not observed. But significant change was observed for both K3 and L3 from pH 4 to pH 10. There was a slight increase in sorption capacity for F3 at pH 4 due to the presence of long aliphatic chains. In case of K3, a decrease of sorption was found in acidic medium due to slight dissociation of HA at lower pH values, the charges surrounding its surface being located in positions less likely to create steric hindrance and/or the association of the organic chemicals with colloidal (soluble HA) in the aqueous phase. Hydrogen bond formation and van-der-Waals attraction forces are considered the main predominant mechanisms of sorption of phenols onto HA.

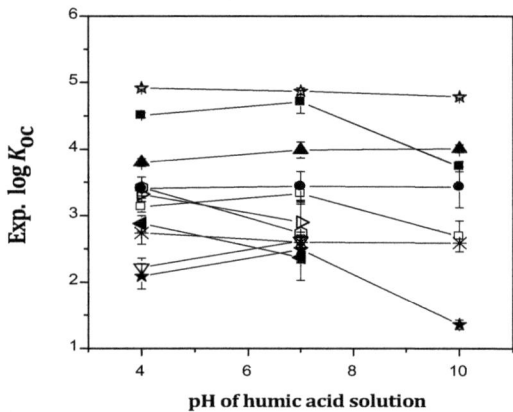

Figure 4.2.6 Exp. log K_{oc} versus pH of HA for acids.
—✳— 2,4-Dichlorophenol, —◀— 2,4,6-Trichlorophenol, —▷— 2,3,4,6-Tetrachlorophenol
—⊙— Pentachlorophenol, —✯— 4-n-Nonylphenol, —□— 3-Phenylphenol, —●— 4,4`-Isopropylidinediphenol
—▲— 2,4-Di-t-butylphenol, —▽— 4-Nitrophenol, —★— Phenol, —■— Triclosan

Figure 4.2.7 showed the effect of pH variation on experimental log K_{oc} values of bases and triazines. The data for M4 and N2 were not detected at pH 4 due to the very high deprotonation (the neutral species were not enough for the uptake by the SPME-fibre and for analysis). It can be seen for G4, H4 and I3, insignificant change was observed between values at pH 7 and pH 4 but significant values were observed between pH 4 and pH 10. Their structures are in neutral form in pH 7, and 10 but speciation can occur in acidic medium, there would be two simultaneous opposite effects of pK_a > pH. Firstly deprotonation of $-NH_3^+$ or $-NH_2^+$- group took place, then the concentration of neutral form of bases decreased and as expected, a decrease in log K_{oc} values occurred which may be also attributed to the very low speciation of HA at pH 4. Secondly, an increase in sorption in pH 4 may be attributed to a high affinity of HA to react with the aromatic structure. In case of PAHs which have a planar structure, the sorption of planar molecules is strongly bound with organic matter more than non-planar molecules due to relatively strong π-π interactions(electron donor - electron acceptor charge transfer) with aromatic moieties of HA (Zhu et al 2004,Liu et al. 2008 and Qu et al. 2008).

Moreover, A1, A2, N4, and P1 displayed a significant change between pH 4 and 10. In an acidic medium, The sorption decreases due to high energy penalty of hydrogen bonding formation with water. The greater sorption for more basic triazines and anilines on HA possessing higher exchange capacity is due to these possible mechanisms: H-bonding, ligand – exchange or charge transfer reaction. Their sorption capacity decreases in acidic medium because of the Coulomb forces as repulsion forces between positive charges of the protonated form (cations) and the positive charges surrounding HA surface, in addition to the low degree of deprotonation of HA. It is known that cations sorbing quickly onto HA may behave as positive counter ions just as organic cations (which, sorbed via a negative site on the organic matter, complexation of the cation molecules with the functional groups of organic matter occurs), that result agrees with Kah et al. 2006 and Kah 2007 who indicates the order of sorption that the cationic species > neutral species > anionic species, generally the sorption of bases faster and stronger than the sorption of acids.

Moreover, the protonated form of the molecules of the bases can be sorbed by electrostatic attraction between the protonated amino group of triazine or anilines and deprotonated carboxyl groups in HA (Kah 2007 and Franco et al. 2008). Then, the non-protonated primary and secondary amino group of triazine can easily interact with carbonyl and phenolic groups of HA by hydrogen bonding or binding with HA through n-π interactions (electron – donor – acceptor complex) or by physical interaction (van der Waals forces and hydrophobic bonding) (Sheng et al. 2005 and Kah et al. 2006). A3 (2, 3, 4, 6-tetrachloroaniline) has an estimated pK_a value of -1.4. Moreover, in the selected range of pH and pK_a, it has a stable form of 2, 3, 5, 6-tetrachloroanilinum ion and its exp. log K_{oc} is not affected by the variation of pH values. The sorption mechanism can occur via the electrostatic attraction forces between the protonated amino group $-NH_3^+$ of 2, 3, 5, 6-

tetrachloroanilinum ion and deprotonated carboxyl group in the HA or via ion exchange reactions (Kah 2007 and Franco et al. 2008).

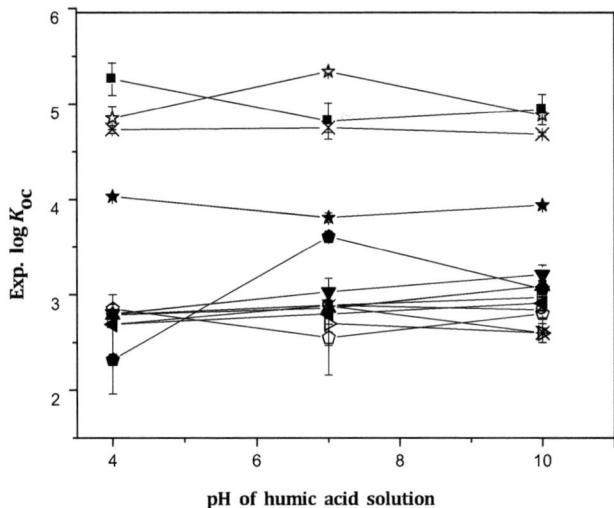

Figure 4.2.7 Exp. log K_{oc} versus pH of HA for bases.

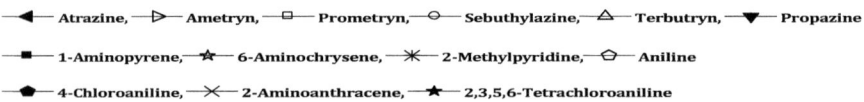

4.2.6 Refitting of LSER-approach for sorption to humic acid

In the beginning of the application of the LSER-approach, the cross-correlation of both experimental and calculated descriptors of the investigated solutes has been checked and shown in Appendix 9.15. The solvent parameters have been calculated using multiple regressions at 95% confidence limit. However, experimental solute descriptors of Abraham and co-workers (Abraham et al. 1987, 1991, 1994, 1999 and 2004) and calculated solute descriptors by increment method (Platts et al. 1999) have been used in the multilinear regressions. Tables 4.2.4 and 4.2.5 show some of the findings that have been observed and will be discussed in detail in the following points:

Firstly, the LSER approach with exp. and calc. input descriptors is applied for total solutes at pH 4, 7 and 10. It appears that the multi-regression of calc. input descriptors have better statistics than exp. input due to calculation availability for all solutes. It has been found that the most positive v term, the deference energy of cavity formation between two phases and most negative b, the hydrogen bond basicity are significant in the sorption process at pH 4. It indicates that the non-ionogenic solutes and neutral acids and deprotonated bases in such pH prefer sorption to HA by lower energy of cavity formation than that in water. Water is more polarized than HA which indicates that solutes prefer water as hydrogen bond donor over HA (hydrogen bonding).

Otherwise at pH 7, and 10 s, a, b, and v are significant. It seems that the ionization of acids and bases at pH 7 and protonated form of bases at pH 10 act as the neutral form of bases, thus dipolarity/polarizability s term is more positive in pH 10 than 7 because of the presence of neutral form of bases in aqueous phase in that form- NH_3^+ or -NH_2^+-. The term a of LSER exp. input is most negative at pH 10 that indicates to the water phase becoming more favorable to be hydrogen bond acceptor over HA. It tends to be the hydrogen bond formation as one of predominant sorption mechanism in addition to electron donor –acceptor transfer.

It generally appears a difference of LSER calc. and exp. input between pH 4 and 10 for both b, and v terms. Moreover, the multilinear regression of LSER calc. input has shown that b and v terms are significant. In contrast, LSER calc. input shows that the most positive v term found at pH 10 indicates that the energy difference (energy penalty) of cavity formation between water and HA is high due to the charged molecules reduces that energy in HA phase especially for bases. Coulomb repulsion force at the surface of HA between the -OH of water and -COOH of HA occurs. In addition, LSER calc. input shows the most negative b term has been found in pH 10 indicating that the water phase is more polarized than HA and solutes prefer water as hydrogen bond donor over HA. LSER calc. input showed better statistical performance than LSER exp. input with low difference between r^2 and q_{cv}^2. LSER exp. inputs are not available for all solutes and the calculated method is better in prediction than experimental method and its difficulty to be determined. The effect of solute structure on sorption mechanism can be investigated in the further details.

Secondly, by applying the LSER exp. and calc. inputs for non-ionogenic solutes, it has been found that **b** and **v**, are only significant in the sorption process for exp. inputs while **e, s, b** and **v** are significant for calc. inputs. These results indicate that the non-ionogenic substances can be sorbed to HA due to various mechanisms such as non-specific intermolecular interactions such as van-der-Waals force and specific interaction as hydrogen bond formation. Moreover, it is now generally recognized that the high positive value of **s**, and **v** term indicates the polarizability of aromatic-rich humic substances and can increase nonspecific interactions such as van-der-Waals interactions between non-ionogenic solutes with aromatic structure and HA due to the formation of charge transfer complex (π-π interactions). It might be reasonable for high energetic of cavity formation in water phase.

In such cases, PAHs act as electron donors and aromatic moieties of HSs act as electron acceptors. For chlorine substitution, the polarizability is more dominant in the sorption process due to the inductive effect of chlorines atoms and their high electronegativity and strong electron withdrawing, where the associated π structures become electron deficient and thus act as π electron acceptors to interact strongly with π-electron donor of polarized π- electron rich HSs (Zhu et al. 2004 and Xing et al. 2008).

Thirdly, when applying the LSER exp. and calc. inputs for acids, it has been found that **b** and **v**, are only significant in the sorption process for exp. inputs while **e, s, a, b** and **v** are significant for calc. inputs. It can be seen from these results that the acids can be sorbed to HA due to various mechanisms such as non-specific intermolecular interactions such as van-der-Waals force and specific interaction as hydrogen bond formation. The negative sign of **a**, and **b** parameters indicates that competition between water molecules and HA to form hydrogen bonds with acidic solutes occurs. The high positive sign of **s** indicates the HA has been interacted with solutes more than water molecules may be attributed to the polarized π- electron rich HSs. It would appear that the water molecules prefer interacting as hydrogen bond donor to acidic solutes more than interacting as hydrogen bond acceptor. Hence, HA prefers to be hydrogen bond acceptor for neutral acids. The high positive value of **v** term indicates that high energy difference of cavity formation of acidic solutes between water and HA due to the less energetic sorption to HA collide due to the charged solutes reduces the energy penalty in water phase and that confirmed with the pervious results.

Table 4.2.4 The multilinear regression statistics of phase parameters using LSER (experimental descriptors by Abraham et al. 1987, 1994a, 1994b, 1999 and 2004).

pH of solution / Substances	e	s	a	b	v	c	r^2	rms	q^2_{cv}	rms_{cv}	N	F-Test
pH 4	0.63 ±0.26	*0.12 ±0.41*	*-0.16 ±0.30*	-1.95 ±0.39	1.62 ±0.29	0.82 ±0.32	0.89	0.31	0.80	0.43	27	32.90
pH 7	*0.05 ± 0.33*	1.16 ±0.51	-0.78 ±0.37	-1.54 ±0.40	1.18 ±0.35	1.08 ±0.37	0.77	0.39	0.62	0.53	29	15.90
pH 10	-0.30 ±0.39	1.59 ±0.72	-1.05 ±0.45	-1.70 ±0.40	1.44 ±0.36	*0.76 ±0.38*	0.81	0.38	0.66	0.53	25	16.03
Non-ionogenic# compounds	*0.14 ±0.24*	*0.06 ±0.62*	....	-2.36 ±0.46	3.01 ±0.39	*-0.20 ±0.27*	0.94	0.24	0.92	0.30	27	93.65
Acids	*-0.06 ±0.34*	0.49 ±0.36	*-0.78 ±0.58*	-0.97 ±0.39	1.89 ±0.26	0.69 ±0.38	0.85	0.29	0.77	0.38	26	22.60
Bases	*0.28 ±0.40*	0.80 ±0.71	-1.42 ±0.66	-1.86 ±0.38	*0.59 ±0.27*	2.40 ±0.40	0.90	0.21	0.86	0.26	28	38.44
Unionized fractions	0.29 ±0.19	0.58 ±0.34	*-0.15 ±0.26*	-1.84 ±0.22	1.50 ±0.21	0.96 ±0.22	0.85	0.34	0.81	0.40	60	60.15
Ionized compounds	*-0.03 ±0.65*	0.98 ±0.75	-1.10 ±0.62	-1.69 ±0.66	1.50 ±0.50	1.10 ±0.51	0.66	0.44	-0.05	0.84	21	5.90

The coefficients e (solvent dispersion interaction), s (the ability of solvent phase to undergo dipole-dipole and dipole-induced dipole interactions with a solute), a (the complementary solvent hydrogen bond basicity), b (the solvent phase hydrogen bond acidity), v (endoergic cavity effect + exoergic solute-solvent effect), and the constant c (solvent specific free energy contribution), are calculated from solutes descriptors via multiparameter linear regression analysis of the dataset. N is no. of points, r^2 is squared correlation coefficient, q^2_{cv} is predictive squared correlation coefficient using leave-one-out-cross validation was calculated as shown in Appendix 9.21 (Schüürmann et al. 2008), rms is root-mean-square error of correlation, rms_{cv} is root-mean-square error of prediction using leave-one-out-cross validation, and F-test was calculated at $\alpha = 0.05$. # means A parameter = zero for all thirteen neutral compounds, thus it was excluded from statistics calculation. Italic format indicates insignificant values at 95% confidence interval. Unionized fractions are unionized fractions of both acids and bases ($1 > f_u > 0.995$) and non-ionogenic compounds in all pH values. Ionized compounds are the acids and bases of $f_u < 0.995$.

Table 4.2.5 The multilinear regression statistics of phase parameters using LSER (calculated descriptors by Platts et al. 1999).

pH of solution / Substances	e	s	a	b	v	c	r^2	rms	q_{cv}^2	rms_{cv}	N	F-Test
pH 4	0.69 ±0.36	0.14 ±0.45	-0.25 ±0.23	-1.75 ±0.29	1.68 ±0.23	0.75 ±0.32	0.88	0.37	0.83	0.47	35	44.03
pH 7	0.08 ±0.43	0.58 ±0.53	-0.61 ±0.27	-1.52 ±0.33	1.63 ±0.27	1.19 ±0.35	0.80	0.45	0.73	0.55	37	25.73
pH 10	-0.11 ±0.40	0.93 ±0.52	-0.20 ±0.35	-1.88 ±0.34	1.83 ±0.25	0.70 ±0.31	0.86	0.40	0.82	0.48	33	34.38
Non-ionogenic compounds	0.58 ±0.21	1.03 ±0.34	2.60 ±1.82	-2.73 ±0.60	1.28 ±0.24	0.36 ±0.15	0.96	0.23	0.94	0.28	39	145.87
Acids	-1.15 ±0.45	1.70 ±0.62	-0.99 ±0.29	-1.54 ±0.41	2.30 ±0.20	0.76 ±0.38	0.90	0.28	0.84	0.39	29	40.98
Bases	0.93 ±0.63	0.35 ±0.66	-1.13 ±0.51	-2.06 ±0.39	0.82 ±0.62	2.10 ±0.28	0.93	0.23	0.91	0.28	37	89.36
Unionized fractions	0.36 ±0.20	0.63 ±0.25	0.37 ±0.20	-2.04 ±0.18	1.59 ±0.14	0.74 ±0.17	0.89	0.35	0.87	0.39	80	123.25
Ionized compounds	-1.15 ±0.88	2.07 ±1.18	-1.72 ±0.38	-2.79 ±0.67	2.44 ±0.42	1.06 ±0.56	0.82	0.40	0.69	0.56	25	16.96

See Table 4.2.5 explanation. A parameter was calculated for all solutes as shown in Appendix 9.14.

Fourthly, when applying the LSER exp. and calc. inputs for bases, it has been found that a, b and v, are only significant in the sorption process for exp. inputs while a, b only are significant for calc. inputs. It can be seen from these results that the sorption of bases to HA is favorable. Bases reduces the energy for cavity formation in water phase, that cations (protonated form of bases are neutral form) sorbed quickly onto HA may behave as positive counter ions just as organic cations (which, sorbed via a negative site on the organic matter, complexation of the cation molecules with the functional groups as – COOH of organic matter occurs). The negative sign of both a, and b accounts for other mechanism that water phase prefers interacting as hydrogen bond donor with basic solutes more than interacting as hydrogen bond acceptor. Then, HA prefers to be hydrogen bond acceptor for deprotonated bases.

Fifthly, the statistical performance of unionized fractions has been improved and more significant after exclusion of the ionized compounds whose $f_u < 0.995$. The predominant mechanism of sorption to organic matter has found as non-specific intermolecular interactions such as van-der-Waals forces as well as specific interaction as hydrogen bond formation.

Sixthly, the sorption of neutral species of ionic compounds of $f_u < 0.995$ can take place via the lower energy of cavity formation in HA phase, and water is preferred to be hydrogen bond donor more than HA. In agreement with previous findings the sorption of charged molecules is not similar to the values mentioned in literature. The order of sorption cationic species > neutral species > anionic species, is now generally recognized and it is approved that the sorption of bases is faster and stronger than the sorption of acids (Kah 2007).

4.2.7 The suggested model to predict log K_{oc} based on pH of for neutral, acidic and basic compounds, and its comparison with the Franco and Trapp model

The theoretical mechanism of the new proposed models of weak acids and bases based on Henderson-Hasselbach equation can be expressed as the following equations (Schüürmann et al. 1998):

$$D_{ow} = f_u K_{ow} + f_i K_{ow} \qquad (4.2.3)$$

D_{ow} is the distribution of the ionogenic compounds between octanol and water, $f_u K_{ow}$ donates the octanol water partition coefficient of the unionized organic species, and $f_i K_{ow}$ donates the sum of both the octanol water partition coefficient of the ionized organic species (K_i) and the distribution coefficient between ion-pairs in octanol and the organic and in organic ions in water (K_{ip})

By combination between equation (4.2.3) with both equations (2.25) and (2.26) from chapter 2, we can calculate D_{ow} for acids and bases. K_{ip} is very small amount, and can be neglected.

$$D_{OW}^{Acids} = \frac{K_{ow}}{1+10^{pH-pKa}} + \frac{K_i}{1+10^{pKa-pH}} \qquad (4.2.4)$$

$$D_{OW}^{Acids} = \frac{K_{ow}}{1+10^{pH-pKa}} + \frac{K_i \cdot 10^{pH-pKa}}{1+10^{pH-pKa}} \qquad (4.2.5)$$

$$D_{OW}^{Acids} = \frac{K_{ow} + K_i . 10^{pH-pKa}}{1+10^{pH-pKa}} \quad (4.2.6)$$

$K_i . 10^{pH-pKa}$ is very small amount and can be neglected.

$$D_{OW}^{Acids} \geq \frac{K_{ow}}{1+10^{pH-pKa}} \equiv f_{u_u}^{Acids} K_{ow} \quad (4.2.7)$$

Then
$$D_{OW}^{Acids} \approx f_{u_u}^{Acids} K_{ow}$$

By taking the logarithmic for both equation sides, then equation (4.2.8) will be obtained.

$$\text{Log } D_{OW}^{Acids} = \log f_{u_u}^{Acids} + \log K_{ow} \quad (4.2.8)$$

In similar way
$$D_{OW}^{Bases} = \frac{K_{ow} + K_i . 10^{pKa-pH}}{1+10^{pKa-pH}} \quad (4.2.9)$$

$K_i . 10^{pKa-pH}$ is very small amount and can be neglected

$$D_{OW}^{Bases} \geq \frac{K_{ow}}{1+10^{pKa-pH}} \equiv f_{u_u}^{Bases} K_{ow} \quad (4.2.10)$$

By taking the logarithmic for both equation sides, then equation (4.2.11) will be obtained

$$\text{Log } D_{OW}^{Bases} = \log f_{u_u}^{Bases} + \log K_{ow} \quad (4.2.11)$$

From equations (4.2.3) and (4.2.4), $\quad D_{ow} = f_u K_{ow} + f_i K_i \quad (4.2.12)$

$$D_{ow}^i = f_i K_{ow}^i \quad (4.2.13)$$

By taking the logarithmic for both equation sides, then equation (4.2.14) will be obtained.

$$\log D_{ow}^i = \log f_i + \log K_{ow}^i \quad (4.2.14)$$

$\log K_{ow}^i$ is very small amount and can be neglected,

Then other form of equation (4.2.3) is equation (4.2.15):

$$D_{ow} = D_{ow}^u + D_{ow}^i \quad (4.2.15)$$

By taking the logarithmic for both equation sides,

$$\text{Log } D_{ow} = a \log D_{ow}^u + b \log D_{ow}^i \quad (4.2.16)$$

By substitutions of equations (4.2.7) and (4.2.13) in equation (4.2.16),

$$\text{Log } D_{ow} = a \, \log (f_u \cdot K_{ow}) + b \log (f_i \, K_i) \quad (4.2.17)$$

$$\text{Log } D_{ow} = a \, (\log f_u + \log K_{ow}) + b \, (\log f_i + \log K_i) \quad (4.2.18)$$

Log f_u and log K_i can be neglected.

Finally the distribution coefficient of acids can be expressed as equation (4.2.19)

$$\text{Log } D_{ow}^{Acids} = a \, \log K_{ow}^{Acids} + b \, \log f_i^{Acids} \quad (4.2.19)$$

The hydrophobicity of acids decreases with increasing pH due to the increase of anionic species, meanwhile the hydrophobicity of bases increases with pH since the neutral form is dominant at pH > pK_a. From the previous equations, It would seem that the hydrophobicity parameter log K_{ow} is important and being a sufficient descriptor for the prediction of the log K_{oc} of neutral compounds. Consideration is made of the important factors like pH and f_u (calculated according to pK_a of the solute). The main objective of our study is to set up models that can be used for polar and non-polar substances. Regarding the proposed models, it should be mentioned that the very low effect of pH variations on the HA structure and very low degree of HA deprotonation at pH 4 can be neglected. Focus was only directed on the variation of pH, on the degree of deprotonation and protonation of solutes according to their pK_a values and their effect on the values of log K_{oc}.

Table 4.2.6 illustrates the proposed model, formula and statistical performance of the new model for acids and bases. Before the setup of suggested model of log K_{oc}-pH dependence for acids and bases, the cross-correlations have been examined. It has been found as follows:

For total acids model $\log K_{ow}$ and (f_i), r^2= 0.0016,

while for total bases $\log f_u K_{ow}$ and $B_{0(est.)}$, r^2 = 0.008

From the previous discussion of LSER approach, it has been found that estimated B_0 was one of the significant parameters in sorption of bases. There is no cross-correlation was found between $B_{0(est.)}$ and hydrophobicity descriptor. Table 4.2.6 showed that the suggested companied model has used a linear equation of two parameters to establish a model of log K_{oc}-pH dependent of weak acids. f_i, the ionized fractions of acids, can be calculated as shown in chapter 2. It has been used and added to hydrophobicities of acids in terms of log K_{ow}. For weak bases, the hydrophobicity of unionized fraction of bases, log f_u K_{ow} can be added to estimated B_0 (hydrogen bond acceptor strength of bases) that has been calculated using increment method by Platts et al. (1999). Outliers of weak base model were 2-aminoanthracene and 1-aminopyrene at pH4 that may be attributed to the sorption enhancement of high hydrophobic PAHs on to sorbent via π-π interactions with aromatic moieties of HA as well as deprotonation of amino group can lower the concentration of protonated form of bases as neutral form. 2,3,5,6-Tetrachloroaniline at pH7 was one of outliers. It has a stable protonation form as 2,3,5,6-tetrachloroanilinum ion because of its pK_a value of -1.4. Hence, its exp. log K_{oc} is not affected by the variation of pH values. The sorption mechanism can occur via the electrostatic attraction forces between the protonated amino group - NH_3^+ of 2,3,5,6-tetrachloroanilinum ion and deprotonated carboxyl group in the HA or via ion exchange reactions, but it's difficult via hydrogen bonding formation.

Table 4.2.6 Results of linear regression of suggested pH-log K_{oc} dependent prediction methods for weak acidic and basic compounds.

Weak acids

Model Equation	pH	N	r^2	rms	q_{cv}^2	rms_{cv}	F-Test	bias	me	mne	mpe
Log $K_{oc(pH)}$ = (0.56 ± 0.06) log K_{ow} – (0.79±0.22) (f_i) + (1.35±0.25)	(4, 7, &10)	29	0.79	0.41	0.75	0.47	48.18	0.00	0.36	-0.80	0.70

J2, J4, K1, and K2 were not detected at pH10.

Weak bases

Model Equation	pH	N	r^2	rms	q_{cv}^2	rms_{cv}	F-Test	bias	me	mne	mpe
Log $K_{oc(pH)}$ = (0.57± 0.05) log $(f_{u(pH)} K_{ow})$ – (1.14 ±0.13) $B_{0(est.)}$+ (2.71±0.19)	(4, 7, &10)	34##	0.79	0.42	0.76	0.46	86.35	0.00	0.25	-0.66	0.82

Outliers were (G4, H4 at pH4, and A3 at pH7), M4 and N2 were not detected at pH4. N is no. of points, r^2 is squared correlation coefficient, q_{cv}^2 is predictive squared correlation coefficient using leave-one-out-cross validation was calculated as shown in Appendix 9.21 (Schüürmann et al. 2008), rms is root-mean-square error of correlation, rms_{cv} is root-mean-square error of prediction using leave-one-out-cross validation, bias is systematic error, me is mean error, mne is maximum negative error, mpe is maximum positive error and F-test was calculated at α = 0.05.

Table 4.2.7 illustrates the comparison between our exp. log K_{oc} data and the calculated data by the Franco and Trapp theoretical model (Franco et al. 2008 and 2009). It was found that low significant correlations and predictive coefficients are associated with high systematic errors. Appeared large difference between r^2 and q^2 has been observed, which indicates that the model did not fit well our experimental findings, especially for bases. The general electrolyte model has the best fitting performance due to a large applicability. It could be applied for all solutes using hydrophobicity descriptor, whereas this model was dependent on the log K_{ow} of neutral species. Appendix 9.18 shows that the limitation of that model for bases at pH4.5. The existing model by Franco and Trapp (2008) does not work for neutral compounds without specification of pK_a and the neutral species of log K_{ow} is required. This is usually not available from experimental data.

Table 4.2.7 Statistical performance of comparison between the experimental log K_{oc} data in the present study and calculated data by Franco and Trapp model (2008).

Model	N	r^2	q^2	rms	F-Test	bias	me	mne	mpe
Electrolyte model	105	0.70	0.53	0.74	160.52	-0.44	0.56	-2.04	0.95
Weak acid model	29	0.69	0.52	0.63	50.24	-0.37	0.49	-1.60	0.41
Weak base model	37	0.44	-0.44	1.11	27.83	-0.85	0.85	-2.32	0.05

See Table 4.2.7 explanation. The equations of model have been shown in Appendix 9.18

Model evaluation and limitations

Our model has limitations for substances which are highly polar and also for strong acids and bases (highly hydrophilic). These limitations arise due to certain problems. Small dataset has been used with simple organic solutes of one functional group. Because of the problem of unavailability of commercial SPME fibre (can uptake the dissociated fractions used in the batch technique), the very small amount of neutral species is quantitatively hard to detect by GC-MSD. The other problem is that the sorption of bases is influenced by many types of forces and mechanisms that can occur simultaneously. Because of Coulomb forces due to the different charges between the sorbent surface and solutes (for instance, ion exchange, ligand exchange and hydrogen bonding), a high sorption capacity of bases is found.

Figure 4.2.8 shows a comparison between different log K_{oc} prediction models for non-ionogenic, acidic and basic solutes. It was found that the LSER approach–calc. input by Platts method shows the best prediction performance with low difference between r^2 and q_{cv}^2. It is important to note that the Abraham descriptors are not available for all solutes; that depends on the molecular descriptors, which can be hard to determine experimentally for more complicated solutes. Therefore, the calc. descriptors for all compounds in the current study are calculated in neutral conditions. Because the exp. descriptors in literature of the subset were determined for compounds in neutral form, the ionization is not being considered. Significant performance of LSER approach –exp input is not detected (in contrary to the previous determination with 61 substances) which may be attributed to narrow dataset in this study (37 substances), and exp. input is not available for all solutes but still needs to be measured. Finally using the LSER approach as prediction method is a pH-independent model, while the proposed prediction models for acids and bases can be used as pH–log K_{oc} based model.

Table 4.2.8 illustrates a comparison of the statistical parameters of different prediction modelling for All compounds and unionized fractions; it has shown an improvement of statistics after exclusion of ionized compounds from dataset (f_u < 0.995). The proposed models in the current study showed the best performance for unionized fractions. The order of model statistics of unionized fractions is: Proposed models are in the current study (non-ionogenic compounds, and acids and bases of f_u< 0.995) > LSER Calc. inputs-in the current study > LSER Exp. inputs-in the current study > classical model (a log K_{ow} +b) > Franco and Trapp > (a log K_{ow} +b pK_a).

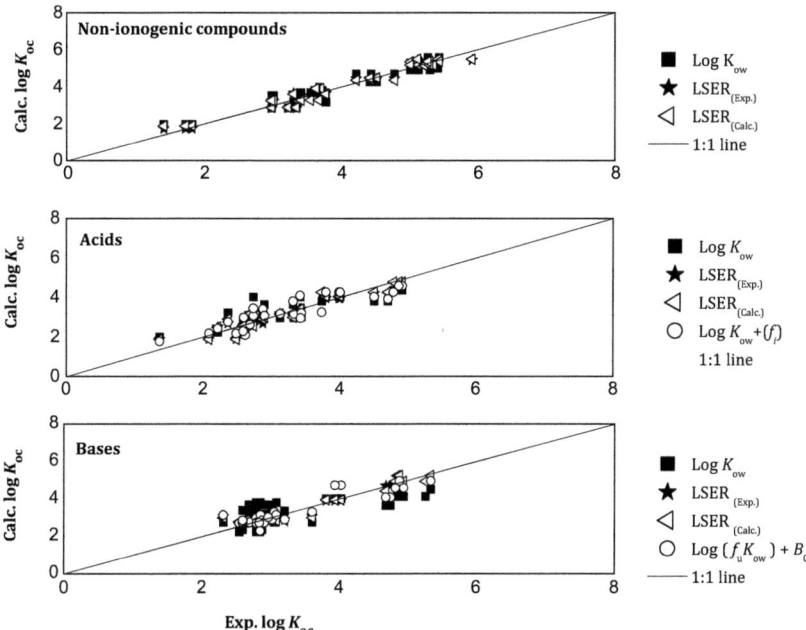

Figure 4.2.8 Results of linear regression of calc. log K_{oc} versus exp. log K_{oc} using different models at 25±2 °C for non-ionogenic compounds, acids and bases.

Table 4.2.8 Statistical parameters of log K_{oc} prediction methods for all compounds and unionized fractions.

Substances	Model	N	r^2	q^2	rms	F-Test	bias	me	mne	mpe
All compounds	Log K_{oc} = a log K_{ow} +b	105	0.69	0.69	0.59	231.53	0.00	0.45	-1.41	1.57
	LSER $_{Exp.}$	81	0.80	0.80	0.39	313.90	0.00	0.32	-0.98	0.85
	LSER $_{Calc.}$	105	0.84	0.84	0.42	551.13	0.00	0.33	-1.07	1.00
	Franco & Trapp	104	0.67	0.45	0.80	146.75	-0.50	0.60	-2.43	0.95
	Log K_{oc} = a log K_{ow} +b pK_a +c	66	0.49	0.49	0.65	60.96	0.00	0.53	-1.40	1.52
	Present Study	101##	0.89	0.89	0.35	799.37	0.00	0.28	-0.80	0.82
Unionized fractions	Log K_{oc} = a log K_{ow} +b	80	0.77	0.77	0.51	264.55	0.00	0.38	-1.36	1.20
	LSER $_{Exp.}$	60	0.85	0.85	0.35	323.03	0.00	0.27	-0.96	0.88
	LSER $_{Calc.}$	80	0.89	0.89	0.35	649.57	0.00	0.27	-0.90	0.79
	Franco & Trapp	80	0.76	0.58	0.70	125.72	-0.43	0.53	-1.79	0.95
	Log K_{oc} = a log K_{ow} +b pK_a +c	41	0.61	0.61	0.55	61.90	0.00	0.45	-1.21	0.86
	Present study	78**	0.92	0.92	0.31	825.97	-0.01	0.25	-0.66	0.80

See Table 4.2.7 explanation. ## Outliers were (I2, G4, H4 at pH4, and A3 at pH7), ** Outliers were (I2 at pH4, A3 at pH7).

4.2.8 Conclusions

The experimental log K_{oc} values of 37 organic compounds have been determined at different pH (4, 7 and 10) and $25 \pm 0.2\,°C$, that data will help in the future studies of bioaccumulation and biodegradation of toxic compounds and assess their risks. The change of pH of HA has no observable effect. The LSER approach has shown that the energy difference of cavity formation between water and HA and hydrogen bonding were the important factors in sorption of polar and non-polar solutes. The sorption of bases to HA is less energetic than that of non-ionogenic and acidic compounds due to their behavior as positive counter ions that interact with negative sites of HA. The sorption of neutral solutes has not been affected by pH variation, while solutes with very high hydrophobicity and aromatic structure show a slight increase of sorption in acidic medium, which can be attributed to π-π interactions with aromatic moieties of HA. The current study introduced a new proposed model under evaluation of log K_{oc}-pH dependence for weak acids and weak bases with full statistics against other existing predictive model by Franco and Trapp (2008) by accompanying log K_{ow} with (f_i of acids) for weak acids and log f_u K_{ow} (f_u of bases) with $B_{0est.}$ (H-Bond acceptor strength) for weak bases predictions. K_{ow} and LSER prediction models for K_{oc} can be improved by excluding ionized compounds from the dataset ($f_u < 0.995$), therefore both modelling methods were fitted and recommended with non-polar compounds more than that of ionogenic behaviour. The batch technique and nd-SPME-GC-MSD are limited with ionic compounds due to speciation, difficult uptake to the fibre and instability of organic acids and bases at high temperature in chromatographic analysis.

4.3 Temperature-dependence of sorption

4.3.1 Introduction

The aim of part 4.3 was to investigate the temperature-dependence of log K_{oc} and calculate the thermodynamic parameters (ΔH_{sor}, ΔS_{sor}, and ΔG_{sor}) controlling the sorption process as well as an attempt to understand the nature of binding with organic matter. There is a lack of literature providing data on how to calculate the sorption thermodynamic parameters based on the mole fraction partition coefficients for chemicals of different functional groups. In addition, the applicability of K_{ow} and LSER approaches to predict of K_{oc} in different temperatures was very interesting. A comparison of different prediction methods was carried out in order to answer some questions raised in literature. Experimental log K_{oc} values were determined for second selected subset of twenty-four compounds from Scheme 3.1 at different temperatures 278.15, 298.15, and 318.15 K, and different types of organic solutes were chosen to reflect a large range of hydrophobicity in terms of log K_{ow} from 0.65 to 8.27 including PAHs, organochlorines, phenols and heterocyclic compounds using batch method combined HS-nd-SPMS-GC-MSD as an analytical tool.

4.3.2 Sorption and temperature effect

Log K_{oc} values of substances were measured in HS-mode as shown in Appendix 9.10, they were calculated from equation (3.6) in chapter 3. Table 4.3.1 shows the experimental values of K_{oc}, K_d, and their logarithmic forms with estimation of maximum logarithmic error of log K_{oc} for all probe compounds. In the beginning of the study, sorption experiments of simetryn and terbuthylazine were carried out at five different temperatures, 278.15, 288.15, 298.15, 308.15, and 318.15 K to explore the best temperature range. As there is no major difference of sorption between 288.15 and 298.15 and between 298.15 and 308.15 K, it has been possible to take only three temperatures covering the scale in order to observe the temperature effect (preliminary screening for optimization of the temperature range). Figure 4.3.1 shows sorption based on mole fraction versus temperature as van't Hoff linear plots (Atkins 1994) for all probe compounds. For ten of these compounds, the plots show a temperature-dependence of sorption, in which log K_{oc} decreases as temperature increases (Tremblay et al. 2005, Niederer et al. 2006, Chen et al. 2007, Jia et al. 2010, and Wang et al. 2011). The other fourteen compounds show an insignificant temperature-dependence of sorption from a statistical point of view, their plots need extra points at least for 5 or 6 different temperatures ($r^2 < 0.98$ for 3 points of data is insignificant).

4.3.3 Thermodynamic parameters based on the mole fraction partition coefficients K_{oc}^x

It was important to explain why in our study the calculations of thermodynamic parameters from log K_{oc}-temperature dependence need to convert the K_{oc} data based on volume concentrations (c_s) to K_{oc} data based on mole fraction. The entropy part (random molecular collisions v) was affected by temperature variation at constant pressure. The differentiation of volume concentrations as variable with respect to temperature as in the following equations (4.3.1), (4.3.2) and (4.3.3) show that it is necessary to calculate (α) the coefficient of thermal (cubical) expansion of the solvent. Thus, it is evident that c_s is not be an independent variable (Guggenhein 1957). Then, the volume concentrations are not convenient in liquid solution

$$\frac{\partial C_s}{\partial T} = -\alpha C_s \qquad (4.3.1)$$

$$\Pi(Cs)\ \Pi(y) = K_c \qquad (4.3.2)$$

y is the activity coefficient and K_c is the equilibrium constant.

$$\frac{\partial \ln K_c}{\partial T} = \frac{\Delta H}{RT^2} - \alpha \Delta v \qquad (4.3.3)$$

R is universal gas constant = 8.3145 JK^{-1}mol^{-1}, T is the absolute temperature in K.

K_{oc}^x the mole fraction partition coefficients of all solutes have been calculated based on the following assumptions in the sorption experiments:

1. It was hypothesized that the HA is represented by the model structure $C_{27}H_{26}N_6O_3$ as suggested by (Wolfram Alpha 2012), and that its molecular weight is 482.534 gmol^{-1}. The total organic carbon fraction of the HA (f_{oc}) is 0.3951 (g/g).
2. The density of water and HA are 0.997 and 1.45 g/mL at 25 °C respectively (Relan et al. 1984). The density change with temperature is small and linear, and it constant at pH 7.
3. In the current study, the log K_{oc} values were determined for neutral species which can be measured by HS-SPME-GC-MSD. The effect of ionization at pH7 was very small (0.97 < f_u < 1)

Chapter 4 - Results and Discussion

according to the Henderson-Hasselbalch equation.

4. The effect of temperature variation on SPME fibres was neglected in log K_{oc} and log K_{oc}^x calculations as the amount of depletion was kept below 5% to prevent the disturbance of equilibrium between the solutes and HA.

5. The K_{oc} values for sorption to HA were converted to mole fraction based partition coefficients using the following equation (4.3.4) (Haftka et al. 2010):

$$K_{oc}^x = \frac{MW_c \cdot N_c}{V_w} \cdot K_{oc} \qquad (4.3.4)$$

N_c is number of carbon atoms (27 carbon atoms) with MW_c = 12.011 g mol^{-1} (Emsley 1998) of the model structure of HA. The molar volume of water (v_w) = 0.01807 L mol^{-1} (Marsh 1987).

6. The thermodynamic parameters of sorption process enthalpy change $\Delta H^x_{w \to HA}$, entropy change $\Delta S^x_{w \to HA}$, and standard Gibbs energy change $\Delta G^x_{w \to HA}$ for compound (i) which sorbed from water to HA were calculated based on the mole fraction partition coefficient K_{oc}^x.

$$\Delta G^x_{w \to HA} = -RT \ln K_{oc}^x = \Delta H^x_{w \to HA} - T\Delta S^x_{w \uparrow HA} \qquad (4.3.5)$$

7. By van 't Hoff linear plot log K_{oc}^x versus reciprocal of absolute temperature, the enthalpy changes ΔH_{sor} and entropy changes ΔS_{sor} can be calculated as follows (Atkins PW 1994 and Chiou CT 2002):

$$\ln K_{oc}^x = \frac{-\Delta H^x_{w \to HA}}{RT} + \frac{\Delta S^x_{w \to HA}}{R} \qquad (4.3.6)$$

The slope of the line when plotting $\ln K_{oc}^x$ versus $\frac{1}{T}$ is equal to $-\Delta H^x_{w \to HA}/R$, and the intercept is equal to $\Delta S^x_{w \to HA}/R$

$$\Delta H^x_{w \to HA} = -slope \cdot R \qquad (4.3.7)$$

$$\Delta S^x_{w \to HA} = intercept \cdot R \qquad (4.3.8)$$

Table 4.3.1 The experimental values of sorption coefficients for different organic substances at pH7±0.20 and at different temperatures (f_{oc} of used Aldrich HA is 0.395 g/g).

No.	Substance	CAS No	Log K_{ow}	T [K]	$K_{d(mean)}$ [L/kg]	Log K_d ± Err	$K_{oc(mean)}$ [L/kg$_{oc}$]	Log K_{oc} ±Err
1	Pyridine	110-86-1	0.65	278.15	8	0.89±0.06	20	1.29±0.06
				298.15	9	0.95±0.07	23	1.35±0.07
				318.15	6	0.82±0.27	17	1.22±0.27
2	Benzaldahyde	100-52-7	1.48	278.15	389	2.59±0.16	982	2.99±0.16
				298.15	217	2.34±0.24	548	2.74±0.24
				318.15	58	1.76±0.46	146	2.17±0.46
3	6-Nitroquinoline	613-50-3	1.84	278.15	1310	3.12±0.21	3308	3.52±0.21
				298.15	644	2.81±0.06	1771	3.25±0.06
				318.15	425	2.63±0.06	1073	3.03±0.06
4	Fluorobenzene	462-06-6	2.27	278.15	64.20	1.81±0.33	195.03	2.29±0.33
				298.15	81.13	1.91±0.13	204.87	2.31±0.13
				318.15	7.68	0.89±0.36	19.40	1.29±0.36
5	4-Chlorophenol	106-48-9	2.39	278.15	442	2.65±0.05	1117	3.05±0.05
				298.15	140	2.15±0.14	917	2.96±0.14
				318.15	113	2.05±0.11	285	2.46±0.11
6	Toluene	108-88-3	2.73	278.15	51	1.71±0.32	128	2.11±0.32
				298.15	107	2.03±0.82	270	2.43±0.82
				318.15	479	2.68±0.30	1209	3.08±0.30
7	Simetryn	1014-70-6	2.80	278.15	930	2.97±0.03	2348	3.37±0.03
				288.15	499	2.70±0.26	1261	3.10±0.26
				298.15	346	2.54±0.14	873	2.94±0.14
				308.15	261	2.42±0.01	659	2.82±0.01
				318.15	160	2.20±0.57	404	2.60±0.57
8	1-Nitronaphthalene	86-57-7	3.19	278.15	1253	3.10±0.08	3165	3.50±0.08
				298.15	312	2.49±0.04	1001	3.00±0.04
				318.15	670	2.83±0.08	1691	3.23±0.08
9	Terbuthylazine	5915-41-3	3.21	278.15	1808	3.26±0.01	4565	3.66±0.01
				288.15	751	2.88±0.26	1897	3.28±0.26
				298.15	498	2.70±0.26	1257	3.10±0.26
				308.15	249	2.40±0.01	630	2.80±0.01
				318.15	188	2.28±0.18	476	2.68±0.18
10	2-Nitrofluorene	607-57-8	3.37	278.15	3004	3.48±0.14	7586	3.88±0.14
				298.15	1039	3.02±0.03	2624	3.42±0.03
				318.15	227	2.36±0.68	573	2.76±0.68

Table 4.3.1 Continued.

No.	Substance	CAS No	Log K_{ow}	T [K]	$K_{d(mean)}$ [L/kg]	Log K_d ± Err	$K_{oc(mean)}$ [L/kg$_{oc}$]	Log K_{oc} ±Err
11	Acridine	260-94-6	3.40	278.15	4977	3.70±0.03	12569	4.10±0.03
				298.15	1701	3.23±0.06	4297	3.63±0.06
				318.15	991	3.00±0.05	2502	3.40±0.05
12	1,2,4-Trimethylbenzene	95-63-6	3.63	278.15	523	2.72±0.10	1321	3.12±0.10
				298.15	945	2.98±0.16	2387	3.38±0.16
				318.15	586	2.77±0.42	1479	3.17±0.42
13	2-Fluorobiphenyl	321-60-8	3.96#	278.15	2757	3.44±0.05	6962	3.84±0.05
				298.15	2549	3.41±0.08	6438	3.81±0.08
				318.15	2759	3.44±0.27	5226	3.72±0.27
14	Fluorene	86-73-7	4.18	278.15	2201	3.34±0.05	5559	3.75±0.05
				298.15	761	2.88±0.13	1922	3.28±0.13
				318.15	648	2.81±0.44	1636	3.21±0.44
15	4-n-Hexylphenol	2446-69-7	4.52#	278.15	4093	3.61±0.08	10336	4.01±0.08
				298.15	3786	3.58±0.13	9560	3.98±0.13
				318.15	1986	3.30±0.16	3881	3.59±0.16
16	1,2,3,4-Tetrachlorobenzene	634-66-2	4.60	278.15	32519	4.51±0.04	82119	4.91±0.04
				298.15	12494	4.10±0.03	31550	4.50±0.03
				318.15	1656	3.22±0.38	4183	3.62±0.38
17	2-Methylanthracene	613-12-7	5.00	278.15	37258	4.57±0.03	94087	4.97±0.03
				298.15	12632	4.10±0.02	31900	4.50±0.02
				318.15	11102	4.05±0.03	25209	4.40±0.03
18	Pentachlorobenzene	608-93-5	5.17	278.15	17737	4.25±0.01	44790	4.65±0.01
				298.15	8314	3.92±0.02	20994	4.32±0.02
				318.15	5048	3.70±0.06	12747	4.11±0.06
19	4-n-Octylphenol	1806-26-4	5.50#	278.15	52208	4.72±0.02	131838	5.12±0.02
				298.15	3750	3.57±0.13	9469	3.98±0.13
				318.15	13806	4.14±0.17	34863	4.54±0.17

Table 4.3.1 Continued.

No.	Substance	CAS No	Log K_{ow}	T [K]	$K_{d(mean)}$ [L/kg]	Log K_d ± Err	$K_{oc(mean)}$ [L/kg$_{oc}$]	Log K_{oc} ±Err
20	2,4`-DDD	53-19-0	5.87#	278.15	209613	5.32±0.10	530665	5.72±0.10
				298.15	147763	5.17±0.02	373139	5.57±0.02
				318.15	99082	5.00±0.07	250840	5.40±0.07
21	2,4`-DDE	3424-82-6	6.00#	278.15	218191	5.34±0.11	562341	5.75±0.11
				298.15	101821	5.01±0.04	257123	5.41±0.04
				318.15	14249	4.15±0.10	36073	4.56±0.10
22	2,4`-DDT	789-02-6	6.79#	278.15	118653	5.07±0.05	299628	5.48±0.05
				298.15	52880	4.72±0.05	133873	5.13±0.05
				318.15	53798	4.73±0.10	136197	5.13±0.10
23	PCB 202	2136-99-4	7.73	278.15	2869189	6.46±0.20	7245428	6.86±0.20
				298.15	567246	5.75±0.05	1432439	6.16±0.05
				318.15	1503449	6.18±0.16	3796589	6.58±0.16
24	PCB 180	35065-29-3	8.27#	278.15	1262571	6.10±0.09	3188310	7.09±0.09
				298.15	702399	5.85±0.05	1773734	6.52±0.05
				318.15	274437	5.44±0.11	693023	5.84±0.11

Err is the maximum logarithmic error of exp. K_{oc} and K_d, it was calculated as shown in Appendix 9.20. (#) Values were estimated by KOWWINTM- Episuite™ v.4 software. The exp. log K_{oc} values of underlined substances was calculated from peak area (Method II in chapter 3) due to the very small peaks of their calibration curves. f_{oc} of used Aldrich HA is 0.395 g/g (see chapter 3).

Chapter 4 - Results and Discussion

The thermodynamic parameters of sorption process such as $\Delta H^x_{w \to HA}$, $\Delta S^x_{w \to HA}$ and $\Delta G^x_{w \to HA}$ were calculated and illustrated in Table 4.3.2. The plots show that ten of the organic solutes exhibit a significant sorption-temperature dependence (B1, N1, B2, O3, O4, H3, M2, C3, E2, and D2), while the other fourteen solutes show an insignificant sorption-temperature dependence from a statistical point of view (M3, B3, J1, G1, B4, H2, D1, L2, C2, H1, L4, E3, E4, and D3) due to the fact that $r^2 < 0.98$ for 3 points of dataset is insignificant according to T-test at confidence interval 95%. The increase in temperature is responsible for a decrease of fiber depletion and shorter equilibrium time. Ten of the probe compounds have a hydrophobic character, and their temperature-dependence behaviour can be easily observed from Figure 4.3.1. Those solutes have sufficient capacity for different types of van-der-Waals interactions including direct and induced dipole-dipole interactions, and hydrogen bonding with hydroxyl and amino groups of hydrophilic part of dissolved HA. Non-polar chemicals interact within themselves and with other substances through London dispersion force (instantaneous dipole-induced dipole) as one of van-der-Waals forces which usually represent the main part of the total interaction force in condensed matter, even though they are generally weaker than hydrogen bonding (insufficient capacity to form hydrogen bonding). Meanwhile, for polar chemicals, the enthalpy-related forces are greater, due to the additional contribution of electrostatic interactions that can donate or accept a hydrogen bond (exothermic contribution of enthalpy) (ten Hulscher et al. 1996). These increase as compounds become less polar. It was known that fact of all significant probe compounds showing temperature-sorption dependence have negative ΔG_{sor} for the compounds of $K_{oc} > 1$, which means that the free energy transfer of sorption is an exergonic process and that it is therefore energetically favourable to be sorbed onto HA. Van't Hoff linear plot shows, if the slope is positive value, then $\Delta H_{sor} < 0$ and sorption will be enthalpy driven. If is the slope is negative, then $\Delta H_{sor} > 0$ and sorption will be decreased due to the interactions with water are more favourable than the interactions with HA. If the intercept is positive, then $\Delta S_{sor} > 0$, means disorder increases upon transfer to HA and the solute solubility decreases due to the water shell around the molecule less higher order than HA shell. In that case the sorption will increase and be entropy driven. If the intercept is negative, then $\Delta S_{sor} < 0$, means the order of HA shell around the molecule increases and the sorption will decrease. Scheme 4.1 shows the results summary of temperature-log K_{oc} data dependence based on mole fraction calculations.

Benzaldahyde (B1) 6-Nitroquinoline (N1) PCB 180 (D2) Simetryn (O3) Terbuthylazine (O4) 2-Nitrofluorene (H3) Acridine (M2) Pentachlorobenzene (C3)	B1, N1, D2, O3, O4, H3, M2, and C3 showed temperature-dependence behaviour with enthalpy-driven forces of sorption (ΔH is negative). They have a high contribution of enthalpy change to the spontaneous free energy of sorption. The plots showed decreasing interactions with HA due to change in water solubility.
Toluene (B2) 2,4`-DDD (E2)	B2 showed inverse sorption temperature-dependence with positive ΔH_{sor} and ΔS_{sor} while E2 showed temperature-dependence of sorption with negative ΔH_{sor}. Both show that behavior with entropy-driven forces of sorption with positive entropy change due to the loss of structured water molecules surrounding the solute molecules (hydrophobic effect) and are inconsistent with an adsorption mechanism with a decrease in the entropy of a solute.

Scheme 4.1 Summary of the results of temperature-log K_{oc} data dependence based on mole fraction calculations.

For pyridine (M3), the temperature-dependence of sorption was insignificant as it showed very small enthalpy changes with high error, therefore it was found that entropy-related forces are dominant in the sorption of pyridine due to highly disorder of HA shell around pyridine upon transfer in HA. At pH 7 ± 0.20, 98.3 % of pyridine is in its unionized form and only 1.7 % is protonated, therefore the lone pairs of nitrogen atom is capable of hydrogen bonding formation, so it prefers to be dissolved in the water phase due to the electrical polarization of water molecules. Meanwhile, its enthalpy changes were not large enough to form hydrogen bonding and showed no enough capacity for enthalpy-related force due to its low hydrophobicity (log K_{ow} = 0.65). In the case of very low K_{ow}, the hydrophobic effect of sorption can occur due to large entropy changes.

For 4-n-octylphenol (L4), it is in neutral form at pH 7 ± 0.20 and its sorption-temperature dependence was insignificant. The enthalpy-related forces are great and strongly contribute (polar chemical behavior) to the sorption process due to the additional electrostatic interactions of the phenolic group.

2-Fluorobiphenyl (D1) shows a reduction of London dispersion forces as one of van-der Waals interactions due to high electronegativity of fluorine. It is weakly susceptible to the fleeting dipoles, has low intramolecular attractive forces, making it to being of hydrophobic/non-polar character. Therefore its enthalpy capacity is not strong enough (low negative ΔH_{sor}), and it also shows entropy-related forces sorption dependence as mentioned in literature (Jia et al. 2010).

2,4`-DDD (E2) shows temperature–dependent of sorption behavior. It has a high hydrophobicity (log K_{ow} = 5.87), therefore repulsion from the water phase and strong sorption to the solid phase are favorable. The hydrophobic effect or hydrophobic bonding is based on entropy-driven forces (ΔS_{sor} >0) because of the reduced mobility of water molecules in the solvation shell of the non-polar solute.

However, the enthalpic contribution of energy transfer was found to be favorable, resulting in a strengthening of water-water hydrogen bonds in the solvation shell. The low mobility of water molecules via hydrogen bonds are limited by building a water "cage" around the hydrocarbon molecule. At higher temperatures, when water molecules become more mobile, this energy gain decreases with decreasing in water solubility. As a result of such entropy-enthalpy compensation, the hydrophobic effect (as measured by the free energy of transfer) is only weakly temperature-dependent and becomes smaller at lower temperature, which leads to "cold denaturation" phenomena, the communal aggregation of the hydrophobic solute molecules which become closer and form a bonding between them, reducing the total area exposed to water.

Toluene (B2) and 1,2,4-trimethylbenzene (B4) have a hydrophobic character of low water solubility, resulting in a high rate of desorption-water partitioning. Mainly, the highly ordered water shell around both solutes is the main reason for this behavior and may be attributed to no enough capacity of hydrogen bonds. In HA-water system a molecule of a hydrophobic solute is enveloped by water, surrounding water molecules enter into an 'ice-like' structure over the greater part of its molecular surface that thermodynamically drives solute molecules out of water. This appears in toluene, which shows an inversely temperature-dependent sorption behavior, both enthalpy and entropy change are positive (ΔH and ΔS > 0) due to the disorder increases upon transfer in HA and the water shell order become less highly order than HA shell and then toluene-water interactions become more favorable than toluene- HA interactions. The positive enthalpy changes (hydrophobic effect i.e. loss of structured water molecules surrounding the solute molecules) result in the enthalpic forces finding it difficult to contribute to the free energy transfer of sorption and needing to form a cavity inside the solid phase, as cavity formation becomes more energetically unfavorable.

PCB 202 (D3) has a highly hydrophobic character (log K_{ow} = 7.73), resulting in repulsion from water phase and very low water solubility. However, it showed a poor linear fit of van't Hoff equation with insignificant sorption-temperature dependence. Chlorine atoms have a high electronegativity that could reduce participation in the London dispersion forces. It was expected that the enthaplic forces capacity would not have been enough to contribute in energy transfer. Moreover, the hydrophobic partitioning is entropy-related forces would have been expected because of the reduced mobility of water molecules in solvation shell of the non-polar solute. We could not ignore the shielding effect (steric effect) of chlorine atoms in both structures of PCB 202 (2,2',3,3',5,5',6,6'-octachlorobiphenyl) shows the four Cl atoms sitting in shielded other Cl atoms and two benzene rings), while PCB 180 (2,2',3,4,4',5,5'-hebtachlorobiphenyl) shows only two Cl atoms which are less shielded due to (C-H bonds) in 6,6' positions).

4.3.4 Comparison with mean log K_{oc} in literature

The exp. log K_{oc} values in the current study which were measured at 298.15 K have been compared with three types of mean values of log K_{oc} in literature as shown in Appendix 9.16. This subset includes eight compounds; no data is available in literature (N1, H3, D1, L2, H1, L4, E2, and E4) as shown in Appendix 9.16. As mentioned in the previous chapter, literature data showed that the sorption behaviour of hydrophobic compound to humic acids is mainly similar to soils and sediments due to the expected similar interaction mechanisms. The experimental values of log K_{oc} in literature depend on the organic carbon content and the functional groups which can participate in interactions with organic solutes.

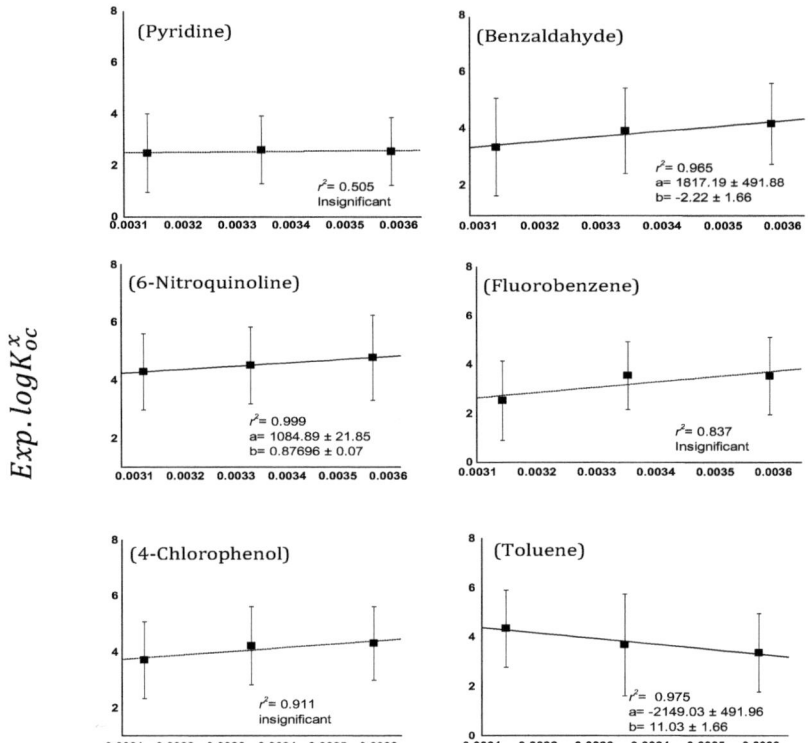

Figure 4.3.1 Van't Hoff plot at different temperatures. r^2 is correlation coefficient, a is slope, b is intercept, straight line is significant correlation and dashed line is insignificant correlation.

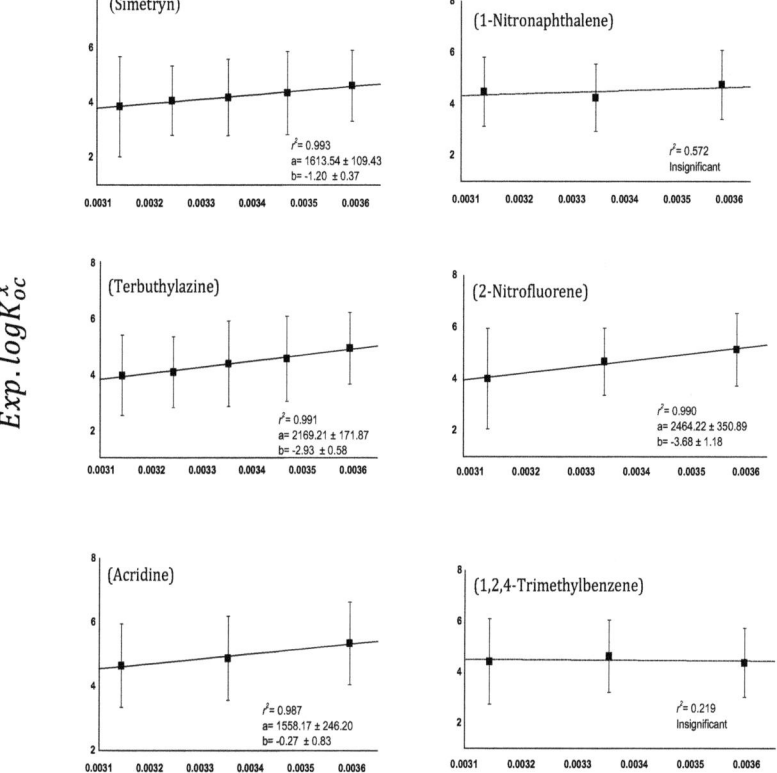

Figure 4.3.1 Continued.

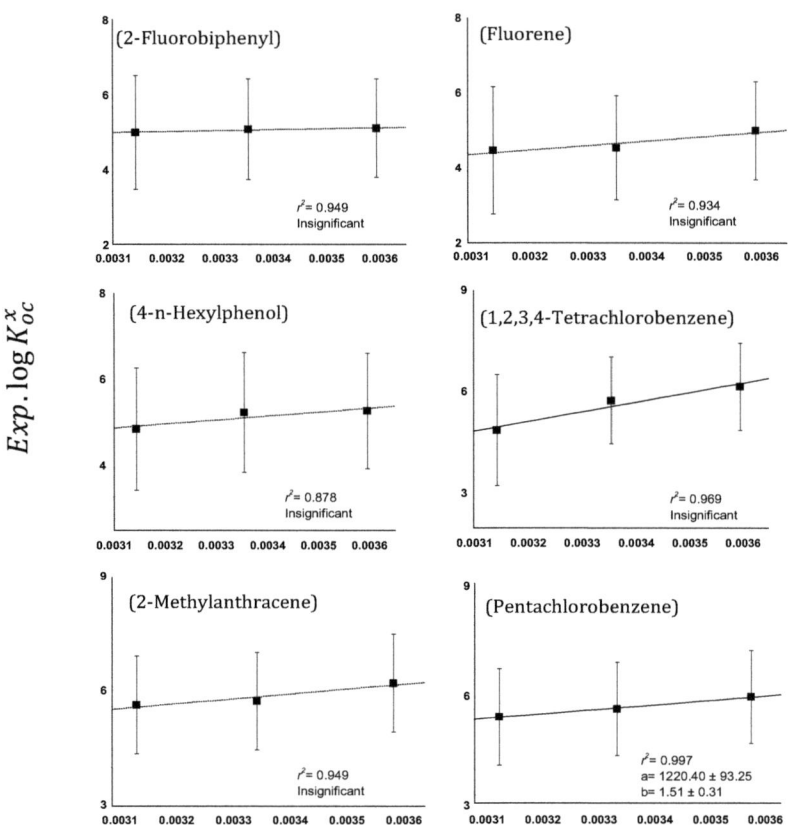

Figure 4.3.1 Continued.

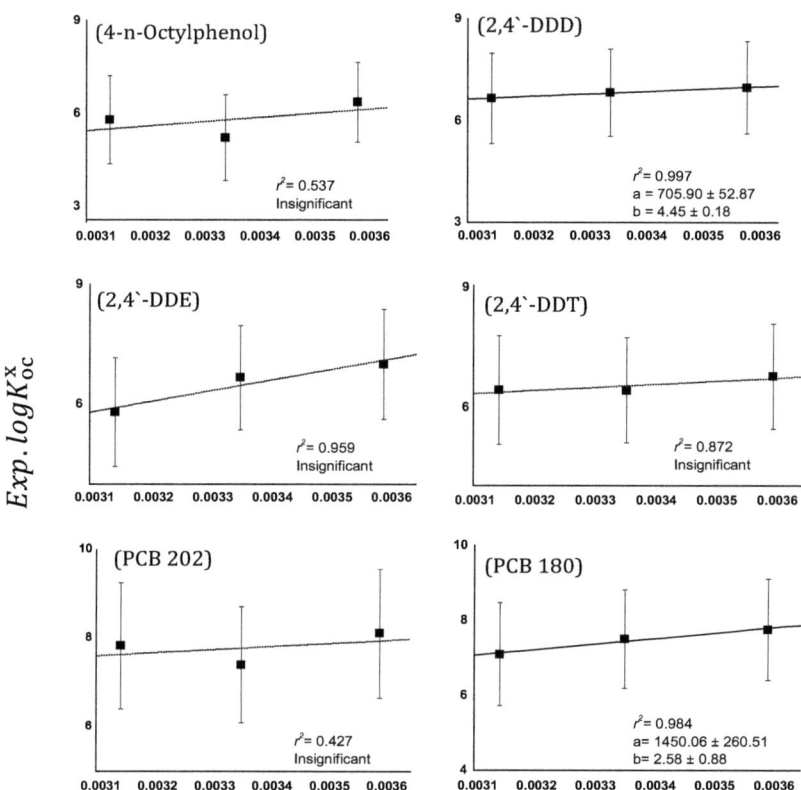

Figure 4.3.1 Continued.

Table 4.3.2 The experimental values of thermodynamic parameters of probe compounds at pH7± 0.20 and at different temperatures.

No.	Substance	T [K]	K_{oc}^{x}	$log K_{oc}^{x}$ ± Err.	ΔG_{sor} kJmol^{-1}	ΔS_{sor} kJ mol^{-1} K^{-1} ± Err.	$-T\Delta S_{sor}$	ΔH_{sor} kJ mol^{-1} ± Err.
1	Pyridine***	278.15	362	2.56 ± 1.32	-13.80	0.040 ± 0.016	-11.01	-2.78 ± 4.75
		298.15	416	2.62 ± 1.32	-14.59		-11.81	
		318.15	308	2.49 ± 1.53	-15.38		-12.60	
2	Benzaldahyde	278.15	17768	4.25 ± 1.42	-22.95	-0.043 ± 0.032	11.84	-34.80 ± 9.42
		298.15	9915	4.00 ± 1.50	-22.10		12.69	
		318.15	2642	3.42 ± 1.72	-21.25		13.54	
3	6-Nitroquinoline	278.15	59853	4.78 ± 1.47	-25.44	0.017 ± 0.001	-4.67	-20.77 ± 0.42
		298.15	32043	4.51 ± 1.32	-25.78		-5.01	
		318.15	19414	4.29 ± 1.31	-26.12		-5.34	
4	Fluorobenzene***	278.15	3529	3.55 ± 1.59	-19.82	-0.077 ± 0.091	21.49	-41.32 ± 27.04
		298.15	3707	3.57 ± 1.38	-18.28		23.04	
		318.15	351	2.55 ± 1.62	-16.73		24.58	
5	4-Chlorophenol***	278.15	20210	4.31 ± 1.31	-23.34	-0.006 ± 0.038	1.65	-24.98 ± 11.24
		298.15	16592	4.22 ± 1.40	-23.22		1.76	
		318.15	5157	3.71 ± 1.37	-23.10		1.88	
6	Toluene	278.15	2316	3.36 ± 1.58	-17.58	0.211 ± 0.032	-58.73	41.15 ± 9.42
		298.15	4885	3.69 ± 2.07	-21.80		-62.95	
		318.15	21875	4.34 ± 1.56	-26.02		-67.17	
7	Simetryn	278.15	42483	4.63 ± 1.29	-24.52	-0.023 ± 0.007	6.38	-30.90 ± 2.10
		288.15	22816	4.36 ± 1.52	-24.29		6.61	
		298.15	15795	4.20 ± 1.40	-24.06		6.84	
		308.15	11924	4.08 ± 1.27	-23.83		7.07	
		318.15	7310	3.86 ± 1.83	-23.60		7.30	
8	1-Nitronaphthalene***	278.15	57265	4.76 ± 1.34	-24.76	0.045 ± 0.059	-12.65	-12.11 ± 17.39
		298.15	18111	4.26 ± 1.30	-25.67		-13.56	
		318.15	30596	4.49 ± 1.34	-26.58		-14.47	

Table 4.3.2 Continued.

No.	Substance	T [K]	K_{oc}^{x}	$\log K_{oc}^{x}$ ± Err.	ΔG_{sor} kJmol^{-1}	ΔS_{sor} kJ mol^{-1} K^{-1} ± Err.	$-T\Delta S_{sor}$	ΔH_{sor} kJ mol^{-1} ± Err.
9	Terbuthylazine	278.15	82596	4.92 ± 1.27	-25.94		15.59	
		288.15	34323	4.54 ± 1.52	-25.38		16.16	
		298.15	22743	4.36 ± 1.52	-24.82	-0.056 ± 0.011	16.72	-41.54 ± 3.29
		308.15	11399	4.06 ± 1.27	-24.26		17.28	
		318.15	8612	3.94 ± 1.44	-23.70		17.84	
10	2-Nitrofluorene	278.15	137256	5.14 ± 1.40	-27.60		19.58	
		298.15	47477	4.68 ± 1.29	-26.20	-0.070 ± 0.023	20.99	-47.19 ± 6.72
		318.15	10367	4.02 ± 1.94	-24.79		22.40	
11	Acridine	278.15	227415	5.36 ± 1.29	-28.39		1.45	
		298.15	77747	4.89 ± 1.32	-28.28	-0.005 ± 0.016	1.55	-29.84 ± 4.71
		318.15	45270	4.66 ± 1.31	-28.18		1.66	
12	1,2,4-Trimethylbenzene***	278.15	23901	4.38 ± 1.36	-23.71		-26.27	
		298.15	43189	4.64 ± 1.42	-25.60	0.094 ± 0.038	-28.16	2.56 ± 11.40
		318.15	26760	4.43 ± 1.68	-27.49		-30.05	
13	2-Fluorobiphenyl***	278.15	125966	5.10 ± 1.31	-27.22		-22.20	
		298.15	116485	5.07 ± 1.34	-28.82	0.080 ± 0.006	-23.80	-5.02 ± 1.66
		318.15	94556	4.98 ± 1.52	-30.41		-25.39	
14	Fluorene***	278.15	100581	5.00 ± 1.31	-26.34		-3.55	
		298.15	34775	4.54 ± 1.39	-26.59	0.013 ± 0.029	-3.81	-22.79 ± 8.66
		318.15	29601	4.47 ± 1.70	-26.85		-4.06	
15	4-n-Hexylphenol***	278.15	187013	5.27 ± 1.34	-28.39		-10.96	
		298.15	172972	5.24 ± 1.39	-29.18	0.039 ± 0.032	-11.75	-17.43 ± 9.48
		318.15	70220	4.85 ± 1.42	-29.97		-12.54	
16	1,2,3,4-Tetrachlorobenzene***	278.15	1485806	6.17 ± 1.29	-33.32		20.80	
		298.15	570845	5.76 ± 1.28	-31.83	-0.075 ± 0.046	22.29	-54.12 ± 13.59
		318.15	75684	4.88 ± 1.64	-30.33		23.79	

Table 4.3.2 Continued.

No.	Substance	T [K]	K_{oc}^{x}	$\log K_{oc}^{x}$ ± Err.	ΔG_{sor} kJmol^{-1}	ΔS_{sor} kJ mol^{-1} K^{-1} ± Err.	-TΔS_{sor}	ΔH_{sor} kJmol^{-1} ± Err.
17	2-Methylanthracene***	278.15	1702347	6.23 ± 1.28	-32.91		-8.45	
		298.15	577177	5.76 ± 1.28	-33.51	0.030 ± 0.027	-9.05	-24.46 ± 8.10
		318.15	456115	5.66 ± 1.29	-34.12		-9.66	
18	Pentachlorobenzene	278.15	810400	5.91 ± 1.27	-31.42		-8.05	
		298.15	379851	5.58 ± 1.28	-32.00	0.029 ± 0.006	-8.63	-23.37 ± 1.79
		318.15	230636	5.36 ± 1.32	-32.57		-9.21	
19	4-n-Octylphenol***	278.15	2385388	6.38 ± 1.28	-32.59		-6.43	-26.16 ± 41.06
		298.15	171326	5.23 ± 1.39	-33.05	0.023 ± 0.138	-6.89	
		318.15	630788	5.80 ± 1.43	-33.52		-7.36	
20	2,4'-DDD	278.15	9601497	6.98 ± 1.36	-37.21		-23.69	
		298.15	6751327	6.83 ± 1.28	-38.91	0.085 ± 0.003	-25.40	-13.52 ± 1.01
		318.15	4538531	6.66 ± 1.33	-40.62		-27.10	
21	2,4'-DDE***	278.15	10174621	7.01 ± 1.37	-37.83		12.43	-50.27 ± 14.66
		298.15	4652211	6.67 ± 1.30	-36.94	-0.045 ± 0.049	13.33	
		318.15	652681	5.81 ± 1.36	-36.04		14.22	
22	2,4'-DDT***	278.15	5421268	6.73 ± 1.31	-35.57		-20.84	
		298.15	2422208	6.38 ± 1.31	-37.06	0.075 ± 0.028	-22.34	-14.72 ± 8.23
		318.15	2464257	6.39 ± 1.36	-38.56		-23.84	
23	PCB 202***	278.15	131093922	8.12 ± 1.45	-42.32		-29.40	
		298.15	25917592	7.41 ± 1.31	-44.43	0.106 ± 0.092	-31.51	-12.92 ± 27.38
		318.15	68692939	7.84 ± 1.42	-46.55		-33.63	
24	PCB 180	278.15	57687146	7.76 ± 1.35	-41.50		-13.73	
		298.15	32092755	7.51 ± 1.31	-42.49	0.049 ± 0.017	-14.72	-27.77 ± 4.99
		318.15	12539094	7.10 ± 1.37	-43.47		-15.71	

1-Insignificant temperature dependence ($r^2 < 0.98$ for 3 points of data) has been shown for substances marked with (***) and in italic format. 2- Insignificant thermodynamic parameters from a statistical point of view are shown in italic format.

4.3.5 Comparison with calculated log K_{oc} from different predictions methods

The exp. log K_{oc} values in the current study (twenty-four compounds as second subset) which were measured at 298.15 K have been compared with the calculated log K_{oc} using different prediction methods as shown in Appendix 9.17. The training set of each method was checked and some methods did not cover all chemical domains which used in the current study as mentioned before in chapter 4.1 and as shown in Appendix 9.22. Overall, the statistics summarized in Table 4.3.3 showed that in linear regression of exp. log K_{oc} versus calc. log K_{oc}, the 2D molecular structure model (Schüürmann et al. 2006) is statistically robust and has the best performance compared to the results achieved with the literature methods. The second best performance was that of the fragment constant model (Tao et al. 1999) with a maximum positive error higher than that of the 2D molecular structure model by 0.32. In addition, the log K_{oc} of benzaldahyde could not be calculated by Tao model. It calculates log K_{oc} from contributions for individual atoms, bonds, functional groups that have been fitted to a data set. Obviously there is no fragment value available for an aldehyde group attached to an aromatic ring. It was found that the training set of both 2D molecular structure and fragment constant models cover most of used organic solutes in this study. The LSER approach by Nguyen et al. (2005) with calc. descriptors input showed the third best performance with a large difference between r^2 and q^2 of 0.24 associated with high *rms* and *mpe* that indicates the limited quality of the calculated Abraham parameters – its original training set was too narrow with regard to the chemical domain and attributed to the presence of systematic error. Franco and Trapp model (2008) (it was applied for only unionized fractions at pH 7 with consideration $pK_a \simeq 50$ for non-polar compounds) showed good regression of 0.20 difference between $r^2 = 0.89$ and $q^2 = 0.69$ with *rms* = 0.67. One of the recent published LSER model by Bronner et al. (2010a) of experimental descriptors showed fit regression $r^2 = 0.90$ but higher than its $q^2 = 0.58$ and this difference due to narrow training set of that model and does not cover all used chemical domain and the presence of the systematic error.

LSER approach by Poole et al. (1999) with calc. descriptors also showed a large difference between r^2 and q^2 by 0.17 but lower than that of the Nugyen model (2005), associated with high *rms*. The LSER approach with experimental input showed for both Nguyen and Poole models lower differences between r^2 and q^2 and lower root-mean-squared-errors than the calculated input descriptors models. The LSER approach of both calc. and exp. input models by Endo et al. (2009) for high and low peat concentrations showed the greatest maximum positive error of all literature methods. It should be kept in mind that for more complex compounds, experimental Abraham descriptors are often not available. For the Poole model, increasing hydrogen bond acidity decreases log K_{oc}, while the opposite is the case with the Nguyen model. The molecular connectivity indices model by Sabljic' et al. (1995) showed low fitting performance with a difference between r^2 and q^2 by 0.16 associated with high error. Other LSER model by Endo et al. (2009a) and Neale et al. (2012) have shown the high difference between high values of r^2 and negative values of q^2 and very high *rms* may be attributed to narrow training dataset and presence of systematic errors.

Table 4.3.3 Results of linear regression of calc. log K_{oc} versus exp. log K_{oc} using different methods in literature.

Model	N	r^2	q^2	rms	F-Test	bias	me	mne	mpe
2D Molecular structure[a]	24	0.92	0.80	0.56	108.96	-0.42	0.44	-1.03	0.15
KOCWIN[b] Molecular topology (top indices +fragments)	24	0.77	0.73	0.65	82.93	-0.21	0.52	-1.70	0.73
KOCWIN[b] (log K_{ow})	24	0.87	0.77	0.60	111.52	-0.37	0.49	-1.23	0.76
Molecular connectivity indices[c]	24	0.83	0.67	0.72	77.84	-0.47	0.58	-1.39	0.77
Fragment constant model[d]	23*	0.91	0.83	0.52	154.24	-0.32	0.40	-1.00	0.47
LSER(Nguyen)[e] exp.	17#	0.89	0.77	0.62	98.91	-0.28	0.49	-1.30	0.46
LSER(Nguyen)[e] calc.	24	0.90	0.66	0.73	123.90	-0.33	0.59	-1.52	0.92
LSER (Poole)[f] exp.	18#	0.86	0.77	0.61	98.34	-0.20	0.49	-0.98	1.36
LSER (Poole)[f] calc.	24	0.88	0.71	0.67	129.76	-0.29	0.55	-1.29	0.94
Trapp model (2008)[g]	22**	0.89	0.69	0.67	56.83	-0.51	0.55	-1.29	0.15
LSER (Endo 2009 – low Peat)[h] exp.	24	0.75	-1.08	1.81	46.01	1.09	1.20	-0.55	4.89
LSER (Endo 2009 – low Peat)[i] calc.	24	0.89	-1.52	2.00	47.82	1.28	1.47	-0.84	4.83
LSER (Endo 2009 – high Peat)[j] exp.	24	0.88	0.27	1.07	87.04	0.27	0.84	-1.45	2.39
LSER (Endo 2009 – high Peat)[k] calc.	24	0.88	-0.05	1.29	75.84	0.10	1.04	-2.09	3.00
LSER (Bronner 2010a)[l] exp.	24	0.90	0.58	0.82	117.09	-0.18	0.67	-1.71	1.21
LSER (Bronner 2010a)[m] calc.	24	0.85	0.06	1.22	72.36	-0.49	0.91	-2.86	1.80
LSER (Neale 2012)[n] exp.	24	0.89	-0.02	1.27	72.97	0.58	1.01	-1.51	2.76
LSER (Neale 2012)[o] calc.	24	0.89	-0.10	1.32	74.20	0.27	1.07	-1.84	3.06

a) Schüürmann et al. 2006. b) KOCWIN software. c) Sabljić et al. 1995. d) Tao et al. 1999. e) Nguyen et al. 2005. f) Poole et al. 1999. g) Franco et al. 2008. h-k) Endo et al. 2009a. l-m) Bronner et al. 2010a. n-o) Neale et al. 2012. # Experimental Abraham descriptors were not available for all solutes (see Appendix 9.13, 9.14, and 9.17). * Log K_{oc} of benzaldehyde could not be calculated. ** The model was used only for 22 unionized fractions at pH 7 ± 0.20. N is no. of points, r^2 is squared correlation coefficient, q^2 is predictive squared correlation coefficient, calculated as shown in Appendix 9.21 (Schüürmann et al. 2008), *rms* is root-mean-square error of correlation, *bias* is systematic error, *me* is mean error, *mne* is maximum negative error, *mpe* is maximum positive error, *F*-test was calculated at α = 0.05.

4.3.6 Refitting of the common K_{ow}-approach for sorption to humic acid

It should be kept in mind that the soil organic carbon content seems to be mainly responsible for the sorption of non-polar organic chemicals which leads to the assumption that the sorption process onto soil can be considered as a partitioning process between water and a hydrophobic soil phase. The hydrophobicity behaviour of non–polar organic chemicals can be expressed as their 1-n-octanol-water partition coefficient (log K_{ow}). The log K_{ow}-log K_{oc} correlations (Karickhoff 1981, Doucette 2003 and Schüürmann et al. 2007) at different temperatures have been performed and their linear regression results with full statistics have been calculated and illustrated in Table 4.3.4.

Table 4.3.4 shows that by raising the temperature, the regression coefficient of hydrophobicity property contribution decreased (slope a), a small difference (0.1) between calibration r^2 and prediction q_{cv}^2 (leave-one-out crossvalidation) was found due to the absence of systematic error. It has been no observable change in regression coefficient with rising temperature. The log K_{ow} and log K_{oc} are both affected by the rising temperature as described in the following equations:

$$\log K_{oc} = a_1 \log K_{ow} + b_1 \quad (4.3.9)$$

By multiplying both sides in 2.303, then

$$\ln K_{oc} = a_2 \log K_{ow} + b_2 \quad (4.3.10)$$

$$\ln K_{oc(T)} = \left(\frac{-\Delta H_{sor}}{RT} + \frac{\Delta S_{sor}}{R}\right) \quad (4.3.11)$$

$$\ln K_{ow(T)} = \frac{\ln K_{oc(T)} - b_{2(T)}}{a_{2(T)}} = \frac{1}{a_{2(T)}}\left(\frac{-\Delta H_{sor}}{RT} + \frac{\Delta S_{sor}}{R}\right) - \frac{b_{2(T)}}{a_{2(T)}} \quad (4.3.12)$$

(a_1 is the slope and b_1 is the intercept of linear plot of log K_{ow} versus log K_{oc}, a_2 = 2.303 a_1, and b_2 = 2.303 b_1)

Partitioning to organic/lipophilic phase should become less favorable as the temperature increases. The best regression with lower errors has been found for room temperature.

Figure 4.3.2 shows that the sorption of organic solutes to organic content of HAs decreases from 278.15 K to 318.15 by 0.35 log units due to decreasing sorption capacity, because at higher temperature the mobility of water molecules will increase so that the removal from water is thermodynamically unfavorable and requires more energy transfer.

Table 4.3.4 Results of linear regression of log K_{oc} versus log K_{ow} at different temperatures.

T [K]	Model	N	r^2	rms	q^2_{cv}	rms_{cv}	F-Test	bias	me	mne	mpe
278.15	Log K_{oc} = (0.70±0.05) log K_{ow} + (1.26±0.23)	24	0.89	0.46	0.88	0.51	180.78	0.00	0.37	-0.97	1.06
298.15	Log K_{oc} = (0.62±0.04) log K_{ow} + (1.27±0.19)	24	0.90	0.38	0.90	0.42	210.56	0.00	0.31	-0.84	0.65
318.15	Log K_{oc} = (0.65±0.05) log K_{ow} + (0.91±0.22)	24	0.89	0.42	0.88	0.48	182.26	0.00	0.34	-0.92	1.10

N is no. of points, r^2 is squared correlation coefficient, q^2_{cv} is predictive squared correlation coefficient using leave-one-out -cross validation, calculated as shown in Appendix 9.21 (Schüürmann et al. 2008), rms is root-mean-square error of correlation, rms_{cv} is root-mean-square error of prediction using leave-one-out cross validation, F-test was calculated at α = 0.05, bias is systematic error, me is mean error, mne is maximum negative error, and mpe is maximum positive error.

At low K_{ow} it has been found that the log K_{oc} at 278.15 and 298.15 K are similar while at higher log K_{ow}, the log K_{oc} at 298.15 and 318.15 K are similar. Therefore, the cooling effect observed at higher K_{ow} (cold denaturation phenomena) for higher hydrophobic character is in contrast with the rising temperature effect observed in low K_{ow} for hydrophobic character. As mentioned before at lower temperatures, the hydrophobic effect was found to be entropy-driven because of the reduced mobility of water molecules in the solvation shell of the non-polar solute. However, the enthalpic component of transfer energy was found to be thermodynamically favorable.

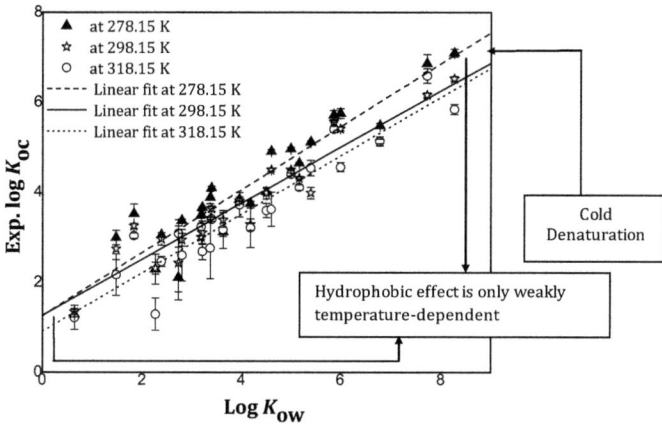

Figure 4.3.2 Exp. log K_{oc} versus log K_{ow} in different temperatures.

4.3.7 Refitting of LSER-approach for sorption to humic acid

In the beginning of the application of the LSER approach, the cross-correlation between experimental and calculated parameters of the investigated dataset has been evaluated in Appendix 9.15. The solvent parameters have been calculated using multiple regressions at 95 % confidence limit at 278.15, 298.15, and 318.15 K. However, experimental descriptors of Abraham and co-workers (Abraham et al. 1987, 1991, 1994, 1999 and 2004), and calculated solute descriptors by increment method (Platts et al. 1999) have been used in the multilinear regressions in the current study. Tables 4.3.5 and 4.3.6 show that the prediction performance of the Platts method using LSER calculated descriptors has been better than that of LSER experimental descriptors. The regression coefficients (r^2) of LSER calc. input was higher than that of LSER exp. input by $\simeq 0.09$, however we could not ignore that experimental descriptors are not available for all probe compounds. LSER exp. input showed higher error with very low F-values of significance. The difference of calibration r^2 and prediction q_{cv}^2 (leave-one–out crossvalidation) is higher for LSER exp. input than LSER calc. input by $\simeq 0.06$-0.10. Large root-mean-squared errors of calibration and prediction using leave-one-out crossvalidation were found for LSER exp. input due to the small size of the data set and unavailability of experimental solute descriptors for all used compounds. It has been found that b and v are significant with higher error for experimental descriptors compared to those calculated. By raising the temperature, v has been increased due to the energy of the cavity formation between water and HA increases due to high energy of cavity formation in water phase. Whereas sorption capacity decreases at high temperature, becoming energetically unfavorable, the mobility of water molecules increases and the removal of solute from water phase to solid becomes more energetically favourable.

For b, a contrary effect has been observed between LSER exp. and LSER calc. inputs as illustrated in Tables 4.3.5 and 4.3.6. The negative value for b indicates that water molecules are better hydrogen bond donors to solutes over HA. The first observed behavior for LSER exp. input was that the negative sign of b increases with temperature, thus indicating that water molecules become better hydrogen bond donors with rising temperature although the high mobility of water molecules at higher temperatures, that is energetically unfavorable to hydrogen bond formation. The contrary behavior was observed for LSER calc. input where negative sign of b decreased, indicating that water molecules do not become better hydrogen bond donors with rising temperature and that HA is a good hydrogen bond donor at higher temperatures.

It has been found that of the dissolved HA interaction parameters, the e, s, a phase parameters are insignificant with great errors for the temperature range used, which means that the sorption of organic solute onto HA has not been affected by the molar refraction, di-polarity and polarisability, and H-bond basicity of HA molecules. The positive value of a indicates that solutes prefer to donate hydrogen atoms to form hydrogen bonds to HA more than water, so that HA is a better hydrogen bond acceptor than water, while with rising temperature this effect becomes smaller and water becomes more favourable to be a good hydrogen bond acceptor due to the high mobility of water molecules. The observation that s becomes more negative with rising temperature indicates that solutes can be more polarised in water phase than in solid phase (at higher temperatures the polarization of solutes reduces the interaction with dissolved HA and reforces the interaction with water molecules). Significance performance of LSER approach –calc. input is better than LSER exp. input (in contrary to the previous determination with 61 substances) may be attributed to narrow dataset in this study (24 substances) and exp. input is not available for all solutes but still needs to be measured.

Table 4.3.5 The multilinear regression statistics of phase parameters using LSER (experimental descriptors by Abraham et al. 1987, 1994a, 1994b, 1999 and 2004).

T [K]	e	s	a	b	v	c	r^2	rms	q_{cv}^2	rms_{cv}	N	F-Test
278.15	0.78 ±0.64	-0.73 ±0.67	0.82 ±1.09	-1.61 ±0.71	2.15 ± 0.62	1.08 ±0.62	0.82	0.63	0.71	0.87	17	10.06
298.15	0.79 ±0.55	-0.84 ±0.57	0.36 ±0.93	-1.78 ±0.61	1.74 ±0.53	1.52 ±0.53	0.82	0.54	0.67	0.79	17	9.83
318.15	0.82 ±0.58	-1.19 ±0.60	0.54 ±0.97	-1.81 ±0.63	2.16 ±0.55	1.09 ±0.55	0.83	0.56	0.71	0.80	17	10.61

Table 4.3.6 The multilinear regression statistics of phase parameters using LSER (calculated descriptors by Platts et al. 1999).

T [K]	e	s	a	b	v	c	r^2	rms	q_{cv}^2	rms_{cv}	N	F-Test
278.15	0.75 ±0.72	-0.12 ±0.86	0.64 ±0.52	-1.52 ±0.54	1.96 ±0.39	0.77 ±0.36	0.91	0.41	0.86	0.55	24	36.98
298.15	0.76 ±0.64	-0.33 ±0.77	0.45 ±0.47	-1.47 ±0.49	1.69 ±0.35	1.12 ±0.32	0.91	0.37	0.86	0.49	24	35.54
318.15	0.70 ±0.66	-0.77 ±0.79	0.29 ±0.48	-1.23 ±0.50	2.13 ±0.36	0.78 ±0.33	0.91	0.38	0.87	0.50	24	37.24

The coefficients e (solvent dispersion interaction), s (the ability of solvent phase to undergo dipole-dipole and dipole-induced dipole interactions with a solute), a (the complementary solvent hydrogen bond basicity), b (the solvent phase hydrogen bond acidity), v (endoergic cavity effect + exogeric solute-solvent effect), and the constant c (solvent specific free energy contribution), are calculated from solutes descriptors via multiparameter linear regression analysis of the dataset. N is no. of points, r^2 is squared correlation coefficient, q_{cv}^2 is predictive squared correlation coefficient using leave-one-out crossvalidation was calculated as shown in Appendix 9.21 (Schüürmann et al. 2008), rms is root-mean-square error of correlation, rms_{cv} is root-mean-square error of prediction using leave-one-out cross validation. F-test calculated at α = 0.05. Italic format indicates insignificant values at 95% confidence interval.

Table 4.3.7 Statistical parameters of log K_{oc} prediction methods at different temperatures.

Absolute Temperature [K]	Model name	r^2	rms	q_{cv}^2	rms_{cv}
278.15	Log K_{ow}	0.89	0.46	0.88	0.51
	LSER exp.	0.82	0.63	0.71	0.87
	LSER calc.	0.91	0.41	0.86	0.54
298.15	Log K_{ow}	0.90	0.38	0.89	0.42
	LSER exp.	0.82	0.54	0.67	0.79
	LSER calc.	0.91	0.38	0.86	0.49
318.15	Log K_{ow}	0.89	0.42	0.88	0.48
	LSER exp.	0.83	0.56	0.71	0.80
	LSER calc.	0.91	0.38	0.87	0.50

For LSER exp. input, N is only 17 of the probe compounds as shown in Appendix 9.13.

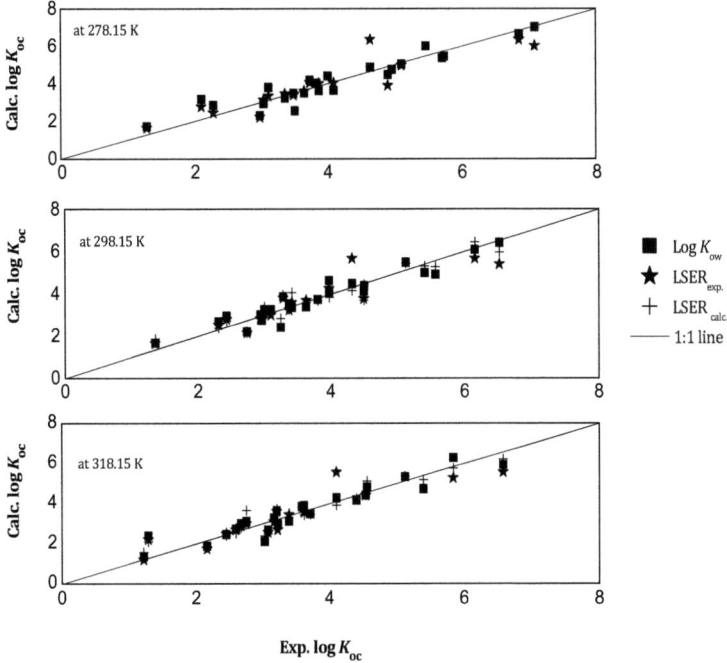

Figure 4.3.3 Results of linear regression of calc. log K_{oc} versus exp. log K_{oc} using different prediction methods at different temperatures.

Figure 4.3.3 shows a comparison of different fitting using predictions methods in literature at 278.15, 298.15, and 318.15 K. Table 4.3.7 shows that the predicted log K_{oc} values from LSER approach with calc. inputs provided the best fitting performance regarding the difference between calibration and prediction statistics, followed by the classical approach of log K_{oc} prediction based on log K_{ow}. The third best performance was observed for LSER exp. input with a difference between r^2 and q^2_{cv} and associated with high rms and rms_{cv}.

4.3.8 Conclusions

The new log K_{oc} data for eight organic compounds (6-nitroquinoline, 2-nitrofluorene, 2-fluorobiphenyl, 4-n-hexylphenol, 2-methylanthracene, 4-n-octylphenol, 2,4´-DDD, and 2,4`-DDT) have been determined at different temperatures (5, 25 and 45 $\pm$ 2 °C). The hydrophobic effect can occur for those compounds of low and high log K_{ow}. Opposite temperature effects for the hydrogen bond acidity ***b*** parameter of HA sorption have been observed between LSER exp. input and LSER calc. input, in addition to the difference in statistics between their r^2 squared correlation coefficient and q_{cv}^2 predictive squared correlation coefficients using leave-one-out cross-validation.

ΔH_{sor}^x, ΔS_{sor}^x, and ΔG_{sor}^x have been calculated from the log K_{oc}^x partition coefficient based on mole fraction - reciprocal of absolute temperature correlations (van 't Hoff plots). A hypothetical model structure of HA has been proposed to be able to estimate the entropy change of the sorption process (spontaneous, exergonic). The log K_{oc} values at different temperatures of ten organic solutes can be predicted using our thermodynamic data. It has been found that eight compounds showed an enthalpy-driven sorption dependency [benzaldahyde, 6-nitroquinoline, simetryn, terbuthylazine, 2-nitrofluorene, acridine, pentachlorobenzene, and PCB 180] while only two showed an entropy-driven sorption dependency [toluene, and 2,4`-DDD]. Measurement limitations of the batch technique and nd-SPME-GC-MSD have been found with ionic compounds due to speciation, difficult uptake to the fibre and instability of organic acids and bases at high temperatures in gas chromatographic analysis. The amount of fibre depletion can be increased by lowering the temperature, therefore it should be kept < 5%.

5. Summary and Conclusions

The sorption of contaminants to dissolved organic matter and sediments affects the transport, distribution and bioavailability of these compounds in the aquatic environment. The concentrations of freely dissolved contaminants can be used to study the structural and thermodynamic behaviour of binding to organic matter. These environmental processes depend on a number of factors such as organic matter concentration, sorption affinity of dissolved organic matter, hydrophobicity and chemical structure of organic contaminant. The site-specific exposure in soil and pore water in terms of freely dissolved contaminant concentrations can be affected by different environmental factors. The determined new log K_{oc} data for some organic compounds (1-aminopyrene, 2,3-dimethylnaphthalene, 2,4`-DDD, 2,4`-DDT, 2,4-di-t-butylphenol, 2-fluorobiphenyl, 2-methylanthracene, 2-nitrofluorene, 3-phenylphenol, 4,4`-isopropylidenediphenol, 4-n-hexylphenol, 4-n-octylphenol, 6-nitroquinoline and sebuthylazine) will help to assess their biodegradation, bioaccumulation and environmental risks.

This thesis focused on the study of the sorption of polar and non-polar compounds in different environmental conditions (i. e. pH, temperature). It also introduces the proposed QSAR modelling with best fitting for experimental data. Chapter 3 is focused on the batch technique and the negligible depletion–SPME method in direct immersion and headspace modes. The first part of chapter 4 covers the study of solute-solvent interactions that are controlled by the chemical structure of both sorbent and sorbate and their physicochemical properties in neutral water. The second and third parts of chapter 4 cover the impact of the variation of the pH of the medium and the equilibrium temperature on the sorption coefficients.

A summary of the main results and conclusions of this thesis is presented in the following paragraphs:

- The solid phase microextraction method (SPME) was applicable for a large diversity of hydrophobic compounds. It is very useful in measuring freely dissolved contaminant concentrations. It allows extracting the freely dissolved amounts from different matrixes in vapour and aqueous phases in a rational way by avoiding the phase separation and clean-up steps before analysis. The fibre depletion should be kept below 5 % to prevent the disturbance of equilibria between the organic solutes and organic matter. Measurement limitations of the batch technique and nd-SPME-GC-MSD have been found with ionic compounds due to speciation, difficult uptake to the fibre and instability of organic acids and bases at high temperature in gas chromatographic analysis. The amount of fibre depletion can be increased by lowering the temperature, therefore it should be kept < 5 %.

- In light of the results of this thesis, it will be expected that future research into the sorption process will show that the sorption capacity of contaminants to soils and sediments is higher than that of humic acids due to the organic carbon content, various composites and the mineral surface of soils which can participate in intermolecular interactions such as hydrogen bond formation. The sorption to soils and sediments will be expected to be more energetically for cavity formation in the soil than that in HA via non-specific interactions.

- This thesis introduces the idea that the prediction of log K_{oc} from log K_{ow} (hydrophobicity property) can be improved by including S, the polarizability of solute, to predict the sorption coefficients of organochlorines and PAHs due to π-π interactions (electron-donor-acceptor interactions). At the same time, the predication of log K_{oc} of phenols was improved after including $A_{est.}$ (H-bond donor strength) to log K_{ow} and a great improvement was found by applying of LSER calc. input only.

- The LSER approach shows differences between two methods of modelling with experimental and calculated input descriptors with a statistical gap between their r^2 squared correlation coefficients and q_{cv}^2 predictive squared correlation coefficients using leave-one-out cross-validation. An improvement of the prediction performance of the K_{ow} and LSER for K_{oc} can be achieved by excluding ionized compounds from the dataset (the unionized fractions f_u < 0.995, and the ionized fractions f_i), therefore both methods of modelling are better fitted and more suitable for non-polar compounds than for the compounds of ionogenic behaviour.

- QSAR modelling based on molecular fragmentation consisting of fragment values and structural correction factors for polar and weakly polar compounds has been shown to provide the most robust statistics for fitting with exp. log K_{oc} data (fragment constant model by Tao et al. 1999 and 2D molecular structure by Schüürmann et al. 2006). This thesis recommends the use of both methods to predict the log K_{oc} values specific for non-polar compounds.

- A new model has been developed to predict the relationship between log K_{oc} and pH for weak acids and weak bases by including (f_i) and $B_{0est.}$ (H-Bond acceptor strength) to log K_{ow} respectively, with full statistical evaluation against other existing predictive models. These proposed models could be applied to a large diverse dataset and guide the researchers to combine the hydrophobicity parameter as a common predictor of log K_{oc} with estimated solute parameters of the LSER approach to obtain the best fitting to the experimental data.

- This thesis introduces an applicable method to calculate the thermodynamic parameters of sorption of organic solutes from water to humic acid through converting the log K_{oc} values to their log K_{oc}^x based on mole fraction using a hypothetical model structure of humic acid. The log K_{oc} data at different temperatures of sorption can be predicted using the thermodynamic relationship to calculate ΔH_{sor}^x, ΔS_{sor}^x and ΔG_{sor}^x for various organic solutes. The results of the thesis have shown that the range of 5 - 45 °C was very representative for the temperature effect, but at least five different temperature points should be chosen to avoid insignificant log K_{oc}-temperature dependence. Polar compounds have shown high errors of log K_{oc} values, which can be attributed to the effect of ionization of organic solutes, in addition to the short equilibrium time by raising temperature that makes the uptake to the fibre more inaccurate and difficult.

- The above mentioned results show that multi-sorption mechanism can be predominant, as had been expected, and that they occur simultaneously due to different types of intermolecular interactions between sorbent and sorbate moieties. Meanwhile, the interactions between the organic solutes and soil organic matter can be carried out through several mechanisms, such as hydrophobic partitioning, hydrogen bonding, electron-donor-acceptor interactions and polar-π interactions, and they can occur simultaneously.

6. Zusammenfassung und Schlussfolgerungen

Die Sorption von Schadstoffen an gelöstem organischem Material und Sedimenten beeinflusst den Transport, die Verteilung und die Bioverfügbarkeit dieser Verbindungen im aquatischen Milieu. Die Konzentrationen von freien gelösten Schadstoffen lassen sich verwenden, um das strukturelle und thermodynamische Verhalten bei der Bindung an organisches Material zu untersuchen. Diese Umweltprozesse hängen von verschiedenen Faktoren ab, wie zum Beispiel der Konzentration des organischen Materials, der Sorptionsaffinität des gelösten organischen Materials, der Hydrophobie und der chemischen Struktur der organischen Schadstoffe. Die regionsspezifische Exposition in Boden und Porenwasser ausgedrückt in der Konzentrationen von freien gelösten Schadstoffen kann von verschiedenen Umweltfaktoren beeinflusst werden. Die ermittelten neuen log K_{oc}-Daten für eine Reihe von organischen Verbindungen (1-Aminopyren, 2,3-Dimethylnaphthalin, 2,4`-DDD, 2,4`-DDT, 2,4-Di-t-Butylphenol, 2-Fluorbiphenyl, 2-Methylanthracen, 2-Nitrofluoren, 3-Phenylphenol, 4,4`-Isopropylidendiphenol, 4-n-Hexylphenol, 4-n-Octylphenol, 6-Nitrochinolin und Sebuthylazin) werden dazu beitragen, ihren Bioabbau, die Bioakkumulation und die Umweltrisiken einzuschätzen.

Der Schwerpunkt dieser Arbeit ist die Untersuchung der Sorption von polaren und nicht-polaren Verbindungen unter verschiedenen Umgebungsbedingungen (z. B. pH, Temperatur). Sie stellt außerdem die vorgeschlagene QSAR-Modellierung mit der besten Anpassung für die experimentellen Daten vor. Kapitel 3 hat als Schwerpunkt die Batch-Methode und die SPME-Methode mit vernachlässigbarer Abreicherung im Direkteintauch- und Headspace-Modus. Der erste Teil von Kapitel 4 beinhaltet die Untersuchung von Solut-Lösungsmittel-Wechselwirkungen, die von der chemischen Struktur sowohl des Sorbents als auch des Sorbats und ihren physikochemischen Eigenschaften in neutralem Wasser kontrolliert werden. Der zweite und dritte Teil von Kapitel 4 befasst sich mit der Wirkung von Veränderungen des pH des Mediums und der Gleichgewichts-Temperatur auf die Sorptionskoeffizienten.

Eine Zusammenfassung der wichtigsten Ergebnisse und Schlussfolgerungen dieser Arbeit wird in den folgenden Absätzen vorgestellt:

- Die Festphasen-Mikroextraktion (SPME) war für eine große Vielfalt von hydrophoben Verbindungen gut anwendbar. Sie ist für die Messung der Konzentrationen von freien gelösten Schadstoffen sehr nützlich. Sie ermöglicht es, die gelösten Mengen aus verschiedenen Matrizen in gasförmigen und wässrigen Phasen zu extrahieren, und vermeidet dabei die Schritte der Phasentrennung und Aufarbeitung vor der Analyse. Die Abreicherung duch die Faser sollte unter 5 % gehalten werden, um die Störung der Gleichgewichte zwischen den organischen Soluten und dem organischen Material zu vermeiden. Bei ionischen Verbindungen wurden Messeinschränkungen bei der Batch-Methode und bei SPME-GCMSD gefunden, aufgrund von Speziation, Aufnahme in die Faser und der Instabilität von Säuren und Basen bei hohen Temperaturen in gaschromatographischen Messungen. Der Umfang der Abreicherung duch die Faser kann durch Erniedrigung der Temperatur erhöht werden, daher sollte er unter 5 % gehalten werden.

- In Anbetracht der Ergebnisse dieser Arbeit ist zu erwarten, dass die zukünftige Erforschung des Sorptionsprozesses zeigen wird, dass aufgrund des Gehalts an organischem Kohlenstoff, verschiedenen Verbindungen und der mineralischen Oberfläche von Böden, die an intermolekularen Wechselwirkungen wie z.B. der Bildung von Wasserstoffbrücken-Bindungen beteiligt sein können, die Schadstoff-Sorptionskapazität von Böden und Sedimenten größer ist als die von Huminsäuren. Es ist zu erwarten, dass die Sorption an Böden und Sedimente für die Hohlraumbildung in Böden energetisch vorteilhafter ist als diese in Huminsäure durch unspezifische Wechselwirkungen.

- Diese Arbeit stelltdie Thesevor, dass die Voraussage von log K_{oc} aus log K_{ow} (Hydrophobie-Eigenschaft) durch die Einbeziehung von S, der Polarisierbarkeit des Soluts, verbessert werden kann, um die Sorptionskoeffizienten von chlororganischen Verbindungen und PAKs unter Berücksichtigung von π-π-Wechselwirkungen vorherzusagen (Elektronen-Donor-Akzeptor Wechselwirkungen). Gleichzeitig wurde eine verbesserte Vorhersage des log K_{oc} von Phenolen durch Einbeziehung von A_{est} (H-Bindungs-Donorstärke) erreicht.

- Der LSER-Ansatz zeigt Unterschiede zwischen den beiden Methoden der Modellierung mit experimentellen und berechneten Deskriptoren, mit einer statistischen Lücke zwischen ihren r^2 quadratischen Korrelationskoeffizienten und q_{cv}^2 prädiktiven quadratischen Korrelationskoeffizienten der Vorhersage unter Verwendung von leave-one-out-Kreuzvalidierung. Eine Verbesserung der statistischen Qualität der K_{ow}- and LSER-Vorhersagemethoden für K_{oc} lässt sich durch den Ausschluss von ionisierten Verbindungen aus dem Datensatz erreichen (die nicht ionisierten Fraktionen $f_u < 0.995$ und die ionisierten Fraktionen f_i), daher sind beide Modellierungsmethoden für nicht-polare Verbindungen besser passend und geeignet als für Verbindungen mit ionogenem Verhalten.

- Es wurde gezeigt, dass die QSAR-Modellierung durch Molekülfragmente, die aus Fragmentwerten und strukturellen Korrekturfaktoren für polare und schwach polare Verbindungen besteht, die robustesten Statistiken für die Anpassung mit experimentellen log K_{oc}-Werten liefert (Tao et al. 1999 und Schüürmann et al. 2006). Diese Arbeit empfiehlt die Verwendung beider Methoden für die Vorhersage der log K_{oc}-Werte insbesondere für nicht-polare Verbindungen.

- Für die Vorhersage des Verhältnisses zwischen log K_{oc} und pH für schwache Säuren und schwache Basen wurde ein neues Modell entwickelt, das (f_i) bzw. B_{0est} (H-Bindungs-Akzeptorstärke) mit einbezieht, mit vollem statistischen Vergleich gegenüber anderen bestehenden Vorhersagemodellen. Dieses vorgeschlagene Modell könnte auf einen großes, diversen Datensatz angewendet werden und Forscher dazu anleiten, den Hydrophobie-Parameter als einen klassischer Schätzer von log K_{oc} mit den geschätzten Solut-Parametern der LSER-Methode zu kombinieren, um die beste Anpassung an die experimentellen Daten zu erhalten.

- Diese Arbeit stellt eine anwendbare Methode zur Berechnung der thermodynamischen Parameter der Sorption von organischen Soluten aus Wasser an Huminsäure vor, mittels Umwandlung der log K_{oc}-Werte in ihre log K_{oc}^x auf Grundlage des Molenbruchs und unter Verwendung einer hypothetischen Modellstruktur für die Huminsäure. Die log K_{oc}-Daten von organischen Soluten bei unterschiedlichen Temperaturen können vorausgesagt werden, indem das vorgeschlagene thermodynamische Modell verwendet wird, um ΔH_{sor}^x, ΔS_{sor}^x und ΔG_{sor}^x für verschiedene organische Solute zu berechnen. Die Ergebnisse dieser Arbeit haben gezeigt, dass der Bereich von 5 – 45 °C sehr repräsentativ für den Temperatureffekt war, jedoch sollten mindestens fünf verschiedene Temperaturpunkte gewählt werden, um eine insignifikante log K_{oc}-Temperatur-Abhängigkeit zu vermeiden. Die polaren Verbindungen zeigten hohe Fehler der log K_{oc}-Werte, was den Effekten dem Ionisierung zugeschrieben werden kann, zusätzlich zu der kurzen Gleichgewichtszeit durch Erhöhung der Temperatur, die die Aufnahme an die Faser ungenauer und schwieriger macht.

- Die oben genannten Ergebnisse zeigen, dass Multi-Sorptions-Mechanismen wie erwartet oft vorherrschend sind, und dass sie simultan auftreten, basierend auf unterschiedlichen Arten von intermolekularen Wechselwirkungen zwischen den funktionellen Gruppen von Sorbent und Sorbat. Gleichzeitig können die Wechselwirkungen zwischen organischen Soluten und organischem Bodenmaterial auf mehreren verschiedenen Mechanismen basieren, wie z. B. Hydrophober Verteilung, Wasserstoffbrücken-Bindung, Elektron-Donor-Akzeptor-Wechselwirkungen und Polar-π-Wechselwirkungen, und diese können simultan auftreten.

7. Future Perspectives

The use of the SPME method to measure the freely dissolved concentration of micropollutants and its combination with batch technique greatly extends the diversity of the set of investigated compounds and is likely to improve the quality of measured log K_{oc} in laboratories. SPME as a validated analytical tool will therefore help to obtain precise data that may contribute to the investigation and understanding of molecular interactions as well as to the selection of precise data for modelling.

However, SPME has only a limited use for measuring ionizable organic compounds such as strong acid and bases due to the difficult uptake to fibre for ionized species. Until now, this method has mainly been used for diverse highly hydrophobic organic compounds, rather than for hydrophilic compounds which has COOH or NH_2 groups. The analytical detection of charged organic molecules presents a substantial challenge. It will be an interesting endeavor to examine the possibilities of applying this method to more anionic and cationic compounds.

More sorption studies are needed in the future to understand the fate and behavior of organic compounds, e. g. regarding the bioavailiability and the degradation in different types of soils and sediments, to improve the risk assessment studies by using a various and large diverse of applicability domain through categorization of new organic chemicals from deferent classes with bi, tri and multifunctional groups. In future studies, it would be important to assess how to apply and extend the LSER approach to a larger set of various sorbents as well as to different representative soils and sediments from the environment. The diversity of molecular properties should be larger, and it should include more polar compounds such as pharmaceuticals and ionizable pesticides. The application of the LSER approach should make it possible to better predict their fate and investigate their ionic and non-ionic molecular interactions.

It is essential that in future research, large diverse datasets should be used to investigate the temperature dependence of sorption at a wider range of temperatures, in addition to the application and development of our proposed experimental model to determine the sorption coefficients at different temperatures based on mole fraction for other types of chemical classes.

8. Citations and Bibliography

Abraham MH, Chadha HS, Whiting GS, Mitchell RC. **1994a**. Hydrogen bonding. 32. An analysis of water-octanol and water-alkane partitioning and the Δ log P parameter of Seiler. *J.Pharm. Sci.* 83: 1085-1100.

Abraham MH, Haftvan JA, Whiting GS, Leo A, Taft RS. **1994b**. Hydrogen Bonding. Part 34. The factors that influence the solubility of gases and vapours in water at 298 K, and a new method for its determination. *J.Chem. Soc. Perkin Trans.* 2: 1777-1791.

Abraham MH, Ibrahim A, Zissimos AM. **2004**. Determination of sets of solute descriptors from chromatographic measurements. *Review. J. of Chromatography A* 1037: 29-47.

Abraham MH, McGowan JC. **1987**. The use of characteristic volumes to measure cavity terms in reversed phase liquid chromatography. *Chromatographia* 23: 243-246.

Abraham MH, Poole CF, Poole SK. **1999**. Classification of stationary phases and other materials by gas chromatography. *Review. J. of Chromatography A* 842: 79-114.

Abraham MH, Whiting GS, Doherty RM, Shuely WJ. **1991**. Hydrogen bonding. Part 13. A new method for the characterization of GLC stationary phases-the laffort data set. *J.Chem. Soc. Perkin Trans* 2: 1451-1460.

Abraham MH. **1993**. Scales of solutes hydrogen-bonding: Their construction and application to physicochemical and biochemical processes. *Chemical Society Reviews:* 73-83.

Atkins PW. **1994**. Physical Chemistry, 4th Ed. *Oxford University Press*, Oxford, UK.

Baker JR, Mihelcic JR, Luehrs DC, Hickey JP. **1997**. Evaluation of estimation methods for organic carbon normalized sorption coefficients. *Water Environment Research* 69(2): 136-145.

Basso EA, Gauze GF, Abraham RJ. **2007**. The prediction of ^1H chemical shifts in amines: a semi-empirical and ab initio investigation. *Magn. Reson. Chem.* 45: 749-759.

Bi E, Schmidt TC, Haderlein SB. **2007**. Environmental factors influencing sorption of heterocyclic aromatic compounds to soil. *Environ. Sci. Technol.* 41: 3172-3178.

Böhme A, Paschke A, Vrbka P, Dohnal V, Schüürmann G. **2008**. Determination of Temperature-Dependent Henry's Law Constant of Four Oxygenated Solutes in Water Using Headspace Solid-Phase Microextraction Technique. *J. Chem. Eng. Data* 53: 2873–2877.

Boule P, Bolte M, Richard C. **1999**. Transformation photoinduced in aquatic media by NO3-/ NO2-, FeIII and humic substances. In: Huzinger O (esd), The Handbook of Environmental Chemistry; vol 2, Part L. *Springer-Verlag*. Berlin. pp: 181-213.

Bronner G, Goss KU. **2010a**. Prediction sorption of pesticides and other multifunctional organic chemicals to soil organic carbon. *Environ. Sci. Technol.* 45: 1313-1319.

Bronner G, Goss KU. **2010b**. Sorption of Organic Chemicals to Soil Organic Matter: Influence of Soil Variability and pH Dependence. *Environ. Sci. Technol.* 45: 1307-1312.

Burke S. Understanding the structure of scientific data. LC.GC Europe onlin supplement, RHM Technology Ltd, High Wycombe, Buckinghamshire, UK. **08. 2007**.
http://chromatographyonline.findanalytichem.com/lcgc/data/articlestandard//lcgceurope/502001/4489/article.pdf.

Chefetz B, Maoz A. **2010**. Sorption of the pharmaceuticals carbamazepine and naproxen to dissolved organic matter: Role of structural fractions. *Water Research* 44: 981-989.

Chefetz B, Xing B. **2009**. Relative role of aliphatic and aromatic moieties as sorption domain for organic compounds: A review. *Environ. Sci. Technol.* 43: 1680-1688.

ChemProp *http://www.ufz.de/index.php?en=6738*

Chen B, Johnson EJ, Chefetz B, Zhu L, Xing B. **2005**. Sorption of polar and nonpolar aromatic organic contaminantes by plant cuticular materials: role of polarity and accessibility. *Environ. Sci. Technol.* 39: 6138-6146.

Chen CS, Tseng SJ. **2007**. Effects of cosolvent and temperature on the desorption of polynuclear aromatic hydrocarbons from contaminated sediments of Chien- Jen river, Taiwan. *Soil Sediment Contam.* 16: 507–521.

Chin YP, Peven CS, Weber Jr WJ. **1988**. Estimating soil/sediment partition coefficients for organic compounds by high performance reverse phase liquid chromatography. *Water Res.* 22(7): 873-881.

Chin YP, Weber Jr WJ. **1989**. Estimating the effects of dispersed organic polymers on the sorption of contaminants by natural solids. I. A predictive thermodynamic humic substance organic solute interaction model. *Environ. Sci. Technol.* 23(8): 978-984.

Chiou CT, Porter PE, Schmedding DW. **1983**. Partition equilibria of non-ionic organic compounds between soil organic matter and water. *Environ. Sci. Technol.* 17 (4): 227-231.

Chiou CT. **2002**. Partition and adsorption of organic contaminants in environmental systems. *John Wiley & Sons, Inc.* Hoboken, USA. pp: 30-50.

Choudhry GG. **1984**. Humic substances, Structural, Photophysical, Photochemical and free radical, Aspects and interaction with environmental chemicals. University of Amsterdam. Network USA. *Cordon and Breach Science Publisher Inc.* pp: 100-131.

Congliang Z, Yan W, Fuan W. **2007**. Determination and Temperature dependence of n-Octanol/Water Partition Coefficients for Seven Sulfonamides from (298.15 to 33.15) K. *Bull Korean Chem. Soc.* 928: 1183-1186.

Converting Henry's law constants. Rolf Sander. **April 2008**. Air Chemistry Department, Max-Planck Institute for Chemistry, Mainz, Germany.
http://dionysos.mpcmainz.mpg.de/~sander/res/henry-conv.html.

Cornelissen G, Gustafsson Ö, Bucheli TD, Jonker MTO, Koelmans A A, van Noort PCM. **2005**. Extensive sorption of organic compounds to black carbon, coal, and kerogen in sediments and soils: Mechanisms and consequences for distribution, bioaccumulation, and biodegradation. *Environ. Sci. Technol.* 39: 6881–6895.

Delle Site A. **2001**. Factors Affecting Sorption of Organic Compounds in Neutral Sorbent/Water Systems and Sorption Coefficients for Selected Pollutants. A Review. *J. Phys. Chem. Ref. Data* 30(1): 187-439.

DIN 32645. **2008-11**. Chemical analysis - Decision limit, detection limit and determination limit

Doerffel K. **1966**. Statistik in der analytischen Chemie. *VEB Deutscher Verlag für Grundstoffindustrie*, Leipzig.

Doucette WJ. **2000**. Soil and sediments sorption coefficients. In: Boethling R S, Mackay D (eds), Handbook of property estimation methods for chemicals: environmental and health sciences. *CRC Press LLC*, USA. pp: 141-190.

Doucette WJ. **2003**. Quantitative structure-activity relationships for predicting soil sediment sorption coefficients for organic chemicals. *Ann. Rev. Environ. Toxic. Chem.*22 (8): 1771-1788.

Emsley J. **1998**. The Elements (Oxford Chemistry Guides). *Oxford Univ. Press Inc.*, New York.

Endo S, Grathwohl P, Haderlein SB, Schmidt TC. **2008**. Compund-spesfic factors influencing sorption nonlinearity in natural organic matter. *Environ. Sci. Technol.* 42: 5897-5903.

Endo S, Grathwohl P, Haderlein SB, Schmidt TC. **2009a**. LFERs for soil organic carbon-water distribution Coefficient (K_{oc}) at environmentally relevant sorbate concentrations. *Environ. Sci. Technol.* 43: 3094-3100.

Endo S, Grathwohl P, Haderlein SB, Schmidt TC. **2009b**. Characterization of sorbate properties of soil organic matter and carbonaceous Geosorbents using n-Alkanes and Cycloalkanes as molecular probes. *Environ. Sci. Technol.* 43: 393-400.

Endo S, Grathwohl P, Haderlein SB, Schmidt TC. **2009c**. Effects of native organic material and water on sorption properties of reference diesel soot. *Environ. Sci. Technol.* 43: 3187-3193.

Endo S. **2008**. Characterization of sorption mechanisms to soil organic phases using molecular probe and polyparamter linear free energy relationship Approaches. *PhD Thesis. Eberhard Karls University*, Tübingen.

Feng L, Han S, Wang L, Wang Z, Zhang Z. **1996**. Determination and estimation of partition properties for phenylthio-carboxylates. *Chemosphere* 32(3): 353-360.

Fiore S, Zanetti MC. **2009**. Sorption of phenols: influence of groundwater pH and of soil organic carbon content. *Am J. of Environ.Sci.* 5(4): 546-554.

Franco A, Fu W, Trapp S. **2009**. Influence of soil pH on the sorption of ionizable chemicals: Modeling Advances. *Environmental Toxicology and Chemistry* 28(3): 458-464.

Franco A, Trapp S. **2008**. Estimation of the soil-water partition coefficient normalized to organic carbon for ionisable organic chemicals. *Environmental Toxicology and Chemistry* 27(10): 1995-2004.

Gauthier TD, Shane EC, Guerin WF. **1986**. Fluorescence quenching method for determining equilibrium-constants for polycyclic aromatic-hydrocarbons binding to dissolved organic materials. *Environ. Sci. Technol.* 20: 1162-1166.

Gawlik BM, Kettrup A, Muntau H. **2000**. Estimation of soil adsorption coefficients of organic compounds by HPLC screening using the second generation of the European references soil set. *Chemosphere* 41: 1337-1347.

Gawlik BM, Sotiriou N, Feicht EA, Schulte-Hostede S, Kettrup A. **1997**. Alternative for the determination of the soil adsorption coefficient, Koc, of non-ionicorganic compounds- A review. *Chemosphere* 34(12): 2525-2551.

Georgi A, Kopinke FD. **2002**. Validation of a modified Flory-Huggins concept for description of hydrophobic organic compound sorption on dissolved organic substances. *Environ.Toxic. Chem.* 21: 1766–1774.

Georgi A. **1998**. Sorption von hydrophoben organischen Verbindungen an gelösten Huminstoffen. *Dissertation in der UFZ*- Leipzig.

Gerstl Z. **1990**. Estimation of organic chemical sorption by soils. *J. Contain Hydrol.* 6: 357-375.

Goss KU. **2005**. Predicting the equilibrium partitioning of organic compounds using just one Linear Solvation Energy Relationship (LSER). *Fluid Phase Equilib* 233: 19-22.

Guggenheim EA. **1957**. Thermodynamics. An Advanced Treatment for Chemists and Physicists. *North-Holland Physics Publ.* 3 rd Ed. pp: 298-320.

Haftka JJH, Govers HAJ, Parsons JR. **2010**. Influence of temperature and origin of dissolved organic matter on the partitioning behavior of polycyclic aromatic hydrocarbons. *Environ Sci Pollut Res.* 17: 1070-1079.

Hawthorne SB, Grabanski CB, Miller DJ. **2006**. Measured partitioning coefficients for parent and alkyl polycyclic aromatic hydrocarbons in 114 historically contaminated sediments: Part1: K_{oc} values. *Environmental Toxicology and Chemistry* 25(11): 2901-2911.

Hong H, Wang L, Han S. **1996**. Prediction adsorption coefficients (K_{oc}) for aromatic compounds by HPLC retention factors (k'). *Chemosphere* 32: 343-351.

Ilani T, Schulz E, Chefetz B. **2005**. Interaction of organic compounds with waste water dissolved organic matter: Role of hydrophobic fractions. *J. Environ. Qual.* 34: 552-562.

Ji L, Wan Y, Zheng S, Zhu D. **2011**. Adsorption of tetracycline and sulfamethoxazole on crop residue-derived ashes: Implication for the relative importance of black carbon to soil sorption. *Environ. Sci. Technol.* 45: 5580-5586.

Jia CX, You C, Pan G. **2010**. Effect of temperature on the sorption and desorption of perfluorooctane sulfonate on humic acid. *J. Environ. Sci.* 22: 355-361.

Jonassen KEN, Nielsen T, Hansen PE. **2003**. The application of high-performance liquid chromatography humic acid columns in determination of K_{oc} of polycyclic aromatic compounds *Environmental Toxicology and Chemistry* 22(4): 741–745.

Jonassen KEN. **2003**. Determination of physico-chemical constants in sorption of polycyclic aromatic compounds to soil organic matter -investigated by use of soil organic matter HPLC columns. *Ph D Thesis. Roskilde University*, Denmark.

Kah M, Brown CD. **2006**. Adsorption of ionisable pesticides in soils. *Rev Environ. Contam Toxicol.* 188: 149-217.

Kah M, Brown CD. **2008**. LogD: Lipophilicity for ionisable compounds. *Chemosphere* 72: 1401-1408.

Kah M. **2007**. Behaviour of ionisable pesticides in soils. PhD Thesis. *University of York*, England.

Kaiser K, Guggenberger G. **2000**. The role of DOM sorption to mineral surfaces in the preservation of organic matter in soils. *Organic Geochemistry* 31: 711-725.

Kamlet M J, Taft RW. **1976**. The Solvatochromic Comparison Method. I. The β-Scale of Solvent Hydrogen-Bond acceptor (HBA) Basicities. *J. of the American Chemical Society* 98(2):377-383.

Karickhoff SW. **1981**. Semi-empirical estimation of sorption of hydrophobic pollutants on natural sediments and soils. *Chemosphere* 10: 833-846.

Katagi T. **2004**. Photodegradation of pesticides on plant and soil surfaces. In: George WW (esd), Rev Environ Contam and Toxicol. 182: 2-77.

Katayama A, Bhula R, Burns GR, Carazo E, Felsot A, Hamilton D, Harris C, Kim YH, Kleter G, Koedel W, Linders J, Peijnenburg JGMW, Sabljic A, Stephenson RG, Racke DK, Rubin B, Tanaka K, Unsworth J, Wauchope RD. **2010**. Bioavailability of Xenobiotics in the Soil Environment. Whitacre DM (ed.) *Reviews of Environmental Contamination and Toxicology* 203: 1-83.

Keiluweit M, Kleber M. **2009**. Molecular-Level Interactions in Soils and Sediments: The Role of Aromatic π-Systems. *Environ. Sci. Technol.* 43: 3421-3429.

Kessler W. **2007**. Multvariate Dataenanalyse. *WILEY-VCH Verlag GmbH & Co. KGaA*, Weinheim, Germany.pp: 89-103.

Kipka U, Di Toro DM. **2011**. A linear solvation energy relationship model of organic chemical partitioning to dissolved organic carbon. *Environmental Toxicology and Chemistry* 30(9): 2023-2029.

Kolb B, Ettre LS. **2006**. Static Headspace Analysis Theory and Practice. *Wiley*. Hoboken, NJ.

Kolb B, Welter C, Bichler C. **1992**. Determination of partition coefficients by automatic equilibrium headspace gas chromatography by vapor phase calibration. *Chromatographia* 34: 235-240.

Kopinke FD, Pörschmann J, Georgi A. **1999**. Application of SPME to study sorption phenomena on dissolved humic organic matter. In: *Pawliszyn J (eds),* Application of solid phase microextraction. *The Royal Society of Chemistry.* Thomas Graham House, Cambridge, UK. pp: 111-127.

Kördel W, Dassenakis M, Lintelmann J, Padberg S. **1997 IUPAC**. The importance of natural organic material for environmental processes in waters and soils (Technical Report). *Pure & Appl. Chem.* 69 (7): 1571-1600.

Krop HB, van Noort PCM, Govers HAJ. **2001**. Determination and theoretical Aspects of the equilibrium between dissolved organic matter and hydrophobic organic micropollutants in water (K_{doc}). In: George W W (esd), Rev Environ Contam and Toxicol. 169: 1-122.

Krutz LJ, Senseman SA, Sciumbato AS. **2003**. Solid–phase microextraction for herbicide determination in environmental samples. *J. of chromatography A* 999: 103-121.

Kühne R, Ebert R-U, Schüürmann G. **October 2011**. ChemProp v.5.2.5 Manual Implemented Models. *Department of Ecological Chemistry, Helmholtz Centre for Environmental Research- UFZ. Permoserstr. 15, 04318 Leipzig, Germany:* pp 46-47.

Kühne R, Ebert RU, Schüürmann G. **1997**. Estimation of vapour pressures for hydrocarbons and halogenated hydrocarbons from chemical structure by a neutral network. *Chemosphere* 34: 671-686.

Kühne R, Ebert RU, Schüürmann G. **2005**. Predication of temperature dependency of Henry´s Law constant from chemical structure. *Environ. Sci. Technol.* 39: 6705-6711.

Laor Y, Rebhun M. **2002**. Evidence for nonlinear binding of PAHs to dissolved Humic acids. *Environ. Sci. Technol.* 36(5): 955-961.

Leo A, Hansch C, and Elkins D. **1971**. Partition coefficients and their uses. *Chemical Reviews* 71(6): 525-616.

Liu P, Zhu D, Zheng H, Shi Z, Sun H, Dang F. **2008**. Sorption of polar and nonpolar aromatic compounds to four surface soils of eastern China. *Environmental Pollution* 156: 1053-1060.

Lohninger H. **1994**. Estimation of soil partition coefficients of pesticides from their chemical structure. *Chemosphere* 29(8): 1611-1994.

MacKay AA, Vasudevan D. **2012**. Polyfunctional ionogenic compound sorption: challenges and new approaches to advance predictive models. *Environ. Sci. Technol.* 46: 9209-9223.

Mackenzie K, Georgi A, Kumke M, Kopinke FD. **2002**. Sorption of pyrene to dissolved humic substances and related model polymers. 2. Solid phase micro extraction (SPME) and fluorescence quenching technique (FQT) as analytical methods. *Environ. Sci. Technol.* 36: 4403-4409.

Mader B, Goss KU, Eisenreich S. **1997**. Sorption of nonionic, hydrophobic organic chemicals to mineral surfaces. *Environ. Sci. Technol.* 31: 1079-1086.

Marsh KN. **1987**. Recommended reference materials for the realization of physicochemical properties. (ed). *Blackwell Scientific Publications*, Oxford. pp: 25-27.

Meylan WM, Howard PH, Boethling RS. **1992**. Molecular topology/fragment contribution method for prediction soil sorption coefficients. *Environ. Sci. Technol.* 26:1560-1567.

Meylan WM. **2000**. PCKOCWIN 1.66. *Syracuse Research Corporation*, Syracuse, NY.

Meylan WM. **2004**. KOWWIN 1.67. *Syracuse Research Coporation*, Syracuse, NY.

Meylan WM. EPSUITE. Syracuse Research Corporation, Syracuse, NY. http://www.syrres.com/esc/est_soft.htm.

Moore D S. **2004**. The basic practice of statistics. *Freeman W H Company*, New York.

Neale PA, Escher B I, Schäfer AI. **2009**. pH dependence of steroid hormone-organic matter interactions at environmental concentrations. *Sci Total Environ.* 407: 1164-1173.

Neale PA, Escher BI, Goss KU, Endo S. **2012**. Evaluating dissolved organic carbon-water partitioning using polyparameter linear free energy relationships: Implications for the fate of disinfection by-products. *Water Research.* 46: 3637-3645.

Nguyen TH, Goss KU, Ball WL. **2005**. Poly parameter linear free energy relationships for estimating the equilibrium partition of organic compounds between water and the neutral organic matter in soils and sediments. *Critical Review. Environ. Sci. Technol.* 39(4): 913-924.

Niederer C, Schwarzenbach RP, Goss KU. **2007**. Elucidating differences in the sorption properties of 10 humic and fulvic acids for polar and nonpolar organic chemicals. *Environ. Sci. Technol.* 41: 6711-6717.

Niederer C, Goss KU, Schwarzenbach RP. **2006**. Sorption Equilibrium of a Wide Spectrum of Organic Vapors in Leonardite Humic Acid: Modeling of Experimental Data. *Environ. Sci. Technol.* 40(17): 5374–5379.

Ohenbusch G, Kumke UM, Frimmel HF. **2000**. Sorption of phenols to dissolved organic matter investigated by solid phase micro extraction. *Sci Total Environ* 253: 63-74.

Oren A, Chefetz B. **2012**. Successive sorption–desorption cycles of dissolved organic matter in mineral soil matrices. *Geoderma* 189-190: 108-115.

Organization for Economic Co-operation and Development. *2000*. OECD Guideline for Testing of chemicals 106: Adsorption-desorption using a batch equilibrium method. *OECD Environment Directorate*, Paris, France.

Organization for Economic Co-operation and Development. *2001*. OECD Guideline for Testing of chemicals 121: Estimation of the adsorption coefficient (K_{oc}) on soil and on sewage sludge using HPLC. *OECD Environment Directorate*, Paris, France.

Paschke A, Popp P. *1999*. Estimation of hydrophobicity of organic compounds. In: *Pawliszyn J(eds)*, Application of solid phase microextraction. *The Royal Society of Chemistry*. Thomas Graham House, Cambridge, UK. pp: 140-155.

Paschke A, Schüürmann G. *1998*. Octanol/Water-Partitioning OF Four HCH ISOMERS AT 5, 25, AND 45 °C. *Fresenius Envir. Bull.* 7: 258-263.

Pawliszyn J. *2001*. Solid Phase Microextraction, Headspace analysis of food and flavours. In: Rouseff & Cadwallader (eds), Theory and practice. *Kluwer Academic*. New York.

Peijnenburg WJGM. *2004*. Fate of contaminants in soils. In: Doelman P, Eijsackers H (eds), Vital soil function, value and properties. *Elsevier*. Amsterdam, the Netherlands. pp: 245-280.

Platts JA, Butina D, Abraham MH, Hersey A. *1999*. Estimation of molecular linear free energy relationship descriptors using a group contribution approach. *J. Chem. Inf Comput. Sci.* 39(5): 835-845.

Poerschmann J, Kopinke FD and Pawliszyn J. *1997*. Solid phase microextraction for determining the distribution of chemicals in aqueous matrices. *Anal. Chem.* 69: 597-600.

Poole SK, Poole CF. *1999*. Chromatographic models for the sorption of neutral organic compounds by soil from water and air. *J. of Chromatography A* 845: 381-400.

Pussemier L, De Borger R, Cloos P, Van Bladel R. *1989*. Relation between the molecular structure and the adsorption of aryl carbamates, phenylureas and anilide pesticides in soil and model organic sorbents. *Chemosphere* 18(9/10): 1871-1882.

Puzyn T, Leszczynska D, Leszczynski J. *2009*. Toward the Development of „Nano-QSARs": Advances and Challenges. *Small* 5(22): 2494-2509.

Qu X, Liu P, Zhu D. *2008*. Enhanced sorption of polycyclic aromatic hydrocarbons to tetra-alkyl ammonium modified smectites via cation –π interactions. *Environ. Sci. Technol.* 42: 1109-1116.

Relan PS, Girdhar KK, Khanna SS. *1984*. Molecular configuration of compost's humic acid by viscometric studies. *Plant and Soil* 81: 203-208.

Rosta HA, Vinken R, Lenz M, Schäffer A. *2003*. Sorption and dialysis experiments to assess the bending of phenolic Xenobiotics to dissolved organic matter in soil. *Environmental Toxicology and Chemistry* 22 (4): 746-752.

Sabijić A, Güsten H, Verhaa H, Hermens J. *1995*. QSAR modelling of soil sorption. Improvements and systematic of log K_{oc} vs. log K_{ow} correlations. *Chemosphere* 31: 4489-4514.

Sabljić A. *1987*. On the prediction of soil sorption coefficient of organic pollutants from molecular structure: application of molecular topology model. *Environ. Sci Technol.* 21: 358-366.

Schüürmann G, Ebert RU, Chen J, Wang B, Kühne R. **2008**. External validation and prediction employing the predictive squared correlation coefficient- test set activity mean vs training set activity mean. *J.Chem. Inf. Model.* 48: 2140-2145.

Schüürmann G, Ebert RU, Kühne R. **2006**. Prediction of the sorption of organic compounds into soil from molecular structure. *Environ. Sci. Technol.* 40: 7005-7011.

Schüürmann G, Ebert RU, Nendza M, Dearden J C, Paschke A, Kühne R, van Leeuwen K, Vermeire T. **2007**. Prediction fate-related physicochemical properties. In: Risk assessment of chemicals: An Introduction. *Springer Science*. Dordrecht, Netherland. pp: 375-426.

Schüürmann G, Kühne R, Kleint F, Ebert RU, Rothenbacher C, Herth P. Software system for automatic chemical property estimation from molecular structure. **1997**. In: Chen F, Schüürmann G (eds), In Quantitative Structure-Activity Relationships in Environmental Sciences-VII. *SETAC Press: Pensacola*: 93-114.

Schüürmann G, Markert B. **1998**. Ecotoxicology. Ecological fundamentals, chemical exposure, and biological effects. (eds). *John Wiley and Spektrum Akademischer Verlag*, New York, USA. pp: 665-749.

Schwarzenbach RP, Gschwend PM, Imboden DM. **2003**. Environmental Organic Chemistry. *John Wiely & Sons Inc.* USA. pp: 57-448.

Sheng G, Yang Y, Huang M, Yang K. **2005**. Influence of pH on pesticide sorption by soil containing wheat residue derived char. *Environmental Pollution* 134: 457-463.

Shiu WY, Ma KC, Varhani`čkovà D, Mackay D. **1994**. Chlorophenols and alkylphenols: A review and correlation of environmentally relevant properties and fate in an evaluative environment. *Chemosphere* 29(6): 1155-1224.

Sprunger L, Proctor A, Acree Jr WE, Abraham HM. **2007**. Characterization of the sorption of gaseous and organic solutes onto polydimethyl siloxane solid–phase micro extraction surfaces using the Abraham model. *J. of Chromatography A* 1175: 162-173.

Stevenson FJ. **1976**. Organic matter reactions involving pesticides in soil. In: Kaufman DD, Still GG, Paulson GD, Bandal SK (eds), Bound and conjugated pesticides residues. ACS Symposium series 29. Washington, USA. pp: 180-207.

Stevenson FJ. **1982**. Humus Chemistry: genesis, composition, reactions. *Wiley, NewYork*. pp: 443.

Sutton R, Sposito G. **2005**. Molecular structure in soil humic substances: The new view. *Environ. Sci. Technol.* 39(23): 9009-9015.

Syracuse Research Corporation (SRC). Environmental science-Interactive physprop database. *http://www.syrres.com/what-we-do/databaseforms.aspx?id=386*

Szab G, Guczi J, Kördel W, Zsolnay A, Major V, Keresztes P. **1999**. Comparison of different HPLC stationary phases for determination of Soil-water distribution coefficient, K_{oc}, values of organic chemicals in RP-HPLC system. *Chemosphere* 39(3): 431-442.

Tao S, Piao H, Dawson R, Lu X, Hu H. **1999**. Estimation of organic carbon normalized sorption coefficient (K_{oc}) for soils using the fragment constant method. *Environ. Sci. Technol.* 33: 2719-2725.

Ten Hulscher TEM, Cornelissen G. **1996**. Effect of temperature on sorption equilibrium and sorption kinetics of organic micropollutants-a review. *Chemosphere* 32: 609–626.

Ter Laak LT, Durjava M, Struijs J, Hermens JLM. **2005**. Solid phase dosing and sampling techniques to determine partition coefficients of hydrophobic chemicals in complex matrixes. *Environ. Sci. Technol.* 39: 3736-3742.

Ter Laak LT, Terbekke MA, Hermens JLM. **2009**. Dissolved organic matter enhances transport of PAHs to aquatic organisms. *Environ. Sci. Technol.* 43: 7212-7217.

Ter Laak LT. **2005**. Sorption to soil of hydrophobic and ionic organic compounds: measurement and modeling. *PhD Thesis. Utrecht University*, Netherlands.

Tremblay L, Kohl SD, Rice JA Gagne JP. **2005**. Effects of temperature, salinity, and dissolved humic substances on the sorption of polycyclic aromatic hydrocarbons to estuarine particles. *Mar. Chem.* 96: 21–34.

Tülp H, Fenner K, Schwarzenbach RP, Goss KU. **2009**. pH-dependent Sorption of acidic organic chemicals to soil organic matter. *Environ. Sci. Technol.* 43: 9189-9195.

Tülp H, Goss KU, Schwarzenbach RP, Fenner K. **2008**. Experimental determination of LSER parameters for a set of 76 diverse pesticides and pharmaceuticals. *Environ. Sci. Technol.* 42: 2034–2040.

under repeatability conditions - Terms, methods, evaluation.

Vaes J, Ramos EU, Hamwijk C, Holsteijn I, Blaauboer BJ, Seinen W, Verhaa HJM, Hermens JLM. **1997**. Solid phase microextraction as a tool to determine membrane/water partition coefficients in Vitro Systems. *Chem. Res. Toxicol.* 10: 1067-1072.

Vitha M, Carr PW. **2006**. The chemical interpretation and practice of linear salvation energy relationships in chromatography. Review. *J. of Chromatography A* 1126: 143-194.

Von Oepen B, Kördel W, Klein W, Schüürmann G. **1991b**. Predictive QSAR, models for estimating soil sorption coefficients potential and limitation based on dominating process. *Sci. Tot. Environ.* 109/110: 343-354.

Von Oepen B, Kördel W, Klein W. **1991a**. Sorption of nonpolar and polar compounds to soil: processes, measurement and experience with applicability of the modified OECD-guideline 106. *Chemosphere* 22: 285-304.

Voskamp M. **2004**. Untersuchungen zur Sorption an Huminstoffen: Molekulargewicht und Kohlenstoff-Isotopenfraktionierung. *Dissertation in der Universität Leipzig*.

Wang L, Yang Z, Niu J. **2011**. Temperature-dependent sorption of polycyclic aromatic hydrocarbons on natural and treated sediments. *Chemosphere* 82: 895–900.

Wang X, Guo X, Yang Y, Tao S, Xing B. **2011**. Sorption mechanisms of phenanthrene, lindane, and atrazine with various humic acid fractions from a single soil sample. *Environ. Sci. Technol.* 45: 2124-2130.

Wauchopa RD, Savage KE, Koskinen WC. **1983**. Adsorption–desorption equilibrium of herbicides in soil: A thermodynamic perspective. *Weed Sci.* 31: 744-751.

Wijnja H, Pignatello JJ, Malekani K. **2004**. Formation of π-π complexes between phenanthrene and model π-acceptor humic subunits. *J Environ Qual.* 33(1): 265-275.

Wolfram Alpha, Humic acid, sodium salt, chemical formula. **03-2012**. http://www.wolframalpha.com/entities/chemicals/humic_acid_sodium_salt/.

Wong TC, Gao X. *1998*. The temperature dependence and thermodynamics functions of partitioning of substances P Peptides in Dodecylphosphocholine Micelles. *Biopolymers* 45: 395-403.

Xing B, Ning P and Bo P. *2008*. Part IV - sorption of hydrophobic organic contaminants. Humic substances Review. *Environ Sci Pollut Res.* 15: 554-564.

Xing B, Pignatello JJ. *1997*. Dual-mode sorption of low –polarity compounds in glassy poly (vinylchloride) and soil organic matter. *Environ. Sci. Technol.* 31: 792-799.

Yu Z, Huang W, Song J, Qian Y, Peng P. *2006*. Sorption of organic pollutants by marine sediments: Implication for the role of particulate organic matter. *Chemosphere* 65: 2493-2501.

Zhang X, Gobas F-APC. *1995*. A thermodynamic analysis of the relationships between molecular size, hydrophobicity, aqueous solubility and octanol-water partitioning of organic chemicals. *Chemosphere* 31: 3501-3521.

Zhao YH, Zhang XJ, Wen Y, Sun FT, Guo Z, Qin WC, Qin HW, Xu JL, Sheng LX, Abraham MH. *2010*. Toxicity of organic chemicals to tetrahymena pyriformis: Effect of polarity and ionization on toxicity. *Chemosphere* 79: 72–77.

Zhu D, Hyung S, Pignatello JJ, Lee LS. *2004*. Evidence for π-π electron donor –acceptor interactions between π –Donor aromatic compounds and π –acceptors sites in soil organic matter through pH effects on sorption. *Environ. Sci. Technol.* 38: 4361- 4368.

Zhu D, Pignatello JJ. *2005a*. Characterization of aromatic compound sorptive interactions with black carbon (charcoal) assisted by graphite as a model. *Environ. Sci. Technol.* 39: 2033–2041.

Zhu D, Pignatello JJ. *2005b*. A concentration-dependent multi-term linear free energy relationship for sorption of organic compounds to soils based on the hexadecane dilute-solution reference stat. *Environ. Sci. Technol.* 39: 8817-8828.

Zissimos AM, Du CM, Valko K, Bevon C, Reynolds B, Wood J, Tam KY. *2002*. Calculation of Abraham descriptors from experimental data HPLC system evaluation of five different methods of calculation. *J. Chem. Soc. Perkin Trans.* 2: 2001-2010.

9. Appendix

9.1 List of Appendices

Contents

		Page
Appendix 9.1	List of Appendices	178
Appendix 9.2	List of Chemicals	179
Appendix 9.3	List of Buffers and Solvents	183
Appendix 9.4	List of Equipments	184
Appendix 9.5	List of Materials	185
Appendix 9.6	List of Computer Software	186
Appendix 9.7	The Physico-chemical Properties of Probe Compounds	187
Appendix 9.8	FTIR and GC-MSD Tests for the Used Aldrich Humic acid	189
Appendix 9.9	The Identification of Naphthalene in Humic Acid by GC-MSD	192
Appendix 9.10	The Sorption Experiments Conditions	193
Appendix 9.11	The Mass Spectroscopy Detections	196
Appendix 9.12	The Data of GC-MSD Temperature Programs	199
Appendix 9.13	List of Experimental Linear Solvation Energy Relationship (LSER) Descriptors (Abraham Descriptors)	200
Appendix 9.14	List of Linear Solvation Energy Relationship (LSER) Descriptors (Platts Descriptors)	203
Appendix 9.15	Cross-Correlation of LSER Parameters	206
Appendix 9.16	The Experimental Values of log K_{oc} Relevant to Humic acid, Soils and Sediments from Literatures	207
Appendix 9.17	The Calculated Values of log K_{oc} from Different Prediction Methods	218
Appendix 9.18	The Used Equations of Franco and Trapp Model	221
Appendix 9.19	The Derivation of K_{oc} Equation	223
Appendix 9.20	The Statistical Treatment of Experimental K_{oc} Data	227
Appendix 9.21	Statistical Parameters of Model Calibration with their Mathematical Formula	232
Appendix 9.22	Applicability Domain and Statistics of Prediction Methods in Literature	234

9.2 List of Chemicals

No.	Substances	CAS No.	Smiles	M.Wt. g.mol⁻¹	Provided Company	Purity
1	Humic acid, Sodium salt	68131-04-4	....	2000 - 500000	Aldrich	Laboratory use, Tech.
			Anilines			
2	Aniline	62-53-3	Nc1ccccc1	93.13	Merck KGaA	≥ 99.5 % GC
3	4-Chloroaniline	106-47-8	Nc1ccc(cc1)Cl	127.57	Merck KGaA	≥ 99 % GC
4	2,3,5,6-Tetrachloraniline	3481-20-7	Clc1c(c(cc1Cl)Cl)N	230.91	Aldrich	99%
			Benzene &Substituted Benzenes (-CHO, -CH₃,-3(CH₃), -F,and -Cl)			
5	Benzene	71-43-2	c1ccccc1	78.12	Merck KGaA	99.80%
6	Benzaldahyde	100-52-7	c1ccccc1C=O	106.13	Merck KGaA	≥ 99 % GC
7	Toluene	108-88-3	Cc1ccccc1	92.14	Merck KGaA	98%
8	Fluorobenzene	462-06-6	Fc1ccccc1	96.11	Fluka Chemie	≥ 99.7 % GC
9	1,2,4-Trimethylbenzene	95-63-6	Cc1cc(cc(c1)C)C	120.2	Merck KGaA	≥ 98 % GC
10	1,3,5-Trichlorobenzene	108-70-3	Clc1cc(cc(c1)Cl)Cl	181.45	Fluka Chemie	> 99 %GC
11	1,2,3,4-Tetrachlorobenzene	634-66-2	Clc1c(c(c(cc1)Cl)Cl)Cl	215.89	Aldrich	98 % GC
12	Pentachlorobenzene	608-93-5	Clc1c(c(c(c(c1)Cl)Cl)Cl)Cl	250.34	Riedel-de-Haën AG	99.9 % GC
13	Hexachlorobenzene	118-74-1	Clc1c(c(c(c(c1Cl)Cl)Cl)Cl)Cl	284.78	Aldrich	99%
			Biphenyls			
14	2-Fluorobiphenyl	321-60-8	Fc1ccccc1c2ccccc2	172.20	Aldrich	96%
15	PCB 180 (2,2',3,4,4',5,5'-Heptachlorbiphenyl)	35065-29-3	c1c(c(c(c(c1Cl)Cl)Cl)Cl)-c2c(cc(c(c2)Cl)Cl)Cl	395.33	Riedel-de-Haën AG	99.9 % GC

Chapter 9 - Appendix

16	PCB 202 (2,2',3,3',5,5',6,6'-Octachlorobiphenyl)	2136-99-4	c1c(c(c(c1Cl)Cl)Cl)c2c(c(c(c2Cl)Cl)Cl)Cl	429.77	Promochem GmbH	10 µg/ml in acetonitril
			Organochlorines			
17	Aldrin (1,2,3,4,10,10-Hexachlor- 1,4,4a,5,8,8a-hexahydro- 1,4-*endo*- 5,8-*exo*-dimethanonaphthalin)	309-00-2	C1C2C=CC1C3C2C4(Cl)C(Cl)=C(Cl)C3(C4(Cl)Cl)Cl	364.92	Poly Science Gr., Niles, Illinois, USA	99%
18	Lindane (γ-Hexachlorcyclohexan)	58-89-9	C1(Cl)C(Cl)C(Cl)C(Cl)C(Cl)C1Cl	290.83	Merck KGaA	> 99.5 %
19	2,4'-DDD (o,p'-Dichlorodiphenyldichloroethane)	53-19-0	c2(C(C(Cl)Cl)c1ccc(Cl)cc1)cccc2Cl	320.05	Supelco	99%
20	2,4'-DDE (o,p'-Dichlordiphenyldichlorethen)	3424-82-6	c2(C(=C(Cl)Cl)c1ccd(Cl)cc1)cccc2Cl	318.03	Riedel-de-Haën AG	96 % GC
21	2,4'-DDT (o,p'-Dichlordiphenyltrichlorethan)	789-02-6	c2(C(C(Cl)(Cl)Cl)c1ccc(Cl)cc1)cccc2Cl	354.49	Supelco	99%
22	4,4'-DDD (p,p'-Dichlordiphenyldichlorethan)	72-54-8	c2c(Cl)ccc(C(c1ccc(Cl)cc1)C(Cl)Cl)cc2	320.05	Riedel-de-Haën AG	99 % GC
23	4,4'-DDE (p,p'-Dichlordiphenyldichlorethen)	72-55-9	c2c(Cl)ccc(C(c1ccc(Cl)cc1)=C(Cl)Cl)cc2	318.03	Riedel-de-Haën AG	99 % GC
24	4,4'-DDT (p,p'-Dichlordiphenyltrichlorethan)	50-29-3	c2c(Cl)ccc(C(c1ccc(Cl)cc1)C(Cl)(Cl)Cl)cc2	354.49	Institute of organic industrial chemistry Warsaw - Poland	99.8 %
			PAHs & Substituted PAHs			
25	Naphthalene	91-20-3	c2cc1c(ccc1)cc2	128.18	Promochem GmbH	99 % GC
26	1-Nitronaphthalene	86-57-7	c1(cccc2cccccc12)[N+]([=O])[O-]	173.17	Aldrich	98%
27	2,3-Dimethylnaphthalene	581-40-8	Cc1cc2cc(c1C)ccc2	156.23	Fluka Chemie	≥ 95.0 %
28	Anthracene	120-12-7	c2cc1cc3c(c1cc2)ccc3	178.24	Promochem GmbH	99.90%
29	2-Aminoanthracene	613-13-8	Nc3cc2cc1ccccc1cc2cc3	193.25	Aldrich	96%

Chapter 9 - Appendix

30	2-Methylanthracene	613-12-7	Cc3ccc2cc1ccccc1cc2c3	192.26	Aldrich	97%
31	Fluorene	86-73-7	C2c1ccccc1c3c2ccc3	166.22	Fluka Chemie	≥ 99 % HPLC
32	2-Nitrofluorene	607-57-8	[O-][N+](c2cc1Cc3c(c1cc2)ccc3)=O	211.22	Aldrich	98%
33	1-Aminopyrene	1606-67-3	Nc2c4c1c3c(ccc1cc2)ccc3cc4	219.29	Aldrich	97%
34	Fluoranthene	206-44-0	c4cc1c(c2ccc3cccc1c23)cc4	202.26	Riedel-de-Haën AG	98%
35	Benzo(a)pyrene	50-32-8	c5cc4c(c1c2c(ccc3cccc(cc1)c23)c4)cc5	252.32	Aldrich	98%
36	6-Aminochrysene	2642-98-0	Nc4c1ccccc1c3ccc2ccccc2c3c4	243.31	Aldrich	97%

Phenols

37	Phenol	108-95-2	Oc1ccccc1	94.11	Merck KGaA	≥ 99.9 % GC
38	4-Chlorophenol	106-48-9	Oc1ccc(cc1)Cl	128.56	Riedel-de-Haën AG	99 % GC
39	4-Nitrophenol	100-02-7	Oc1ccc([N+](=O)[O-])cc1	139.11	Riedel-de-Haën AG	99.9 % HPLC
40	2,4-Dichlorophenol	120-83-2	Oc1c(cc(cc1)Cl)Cl	163.00	Merck KGaA	≥ 98% GC
41	2,4,6-Trichlorophenol	88-06-2	Oc1c(cc(cc1Cl)Cl)Cl	197.45	Riedel-de-Haën AG	99.8 % GC
42	2,3,4,6-Tetrachlorophenol	58-90-2	c1(O)c(c(cc(c1Cl)Cl)Cl)Cl	231.89	Riedel-de-Haën AG	96.7 % HPLC
43	Pentachlorophenol	87-86-5	Oc1c(c(Cl)c(c(Cl)c1Cl)Cl)Cl	266.34	Sigma	98%
44	2,4-Di-t-butylphenol	96-76-4	c1(c(O)ccc(C(C)(C)C)c1)C(C)(C)C	206.33	Aldrich	99%
45	3-Phenylphenol	580-51-8	Oc2cc(c1ccccc1)ccc2	170.21	Aldrich	90%
46	4,4'-Isopropylidenediphenol	80-05-7	C(C)(C)(c1ccc(O)cc1)c2ccc(O)cc2	228.29	Fluka Chemie	97 % HPLC

- 181 -

#	Name	CAS	SMILES	MW	Supplier	Purity
47	4-n-Hexylphenol	2446-69-7	c1(CCCCCC)ccc(O)c1	178.27	Labor Dr. Ehrenstorfer-Schäfers, Augsburg GmbH Germany	99.50%
48	4-n-Nonylphenol	104-40-5	c1(ccc(CCCCCCCCC)cc1)O	220.36	Labor Dr. Ehrenstorfer-Schäfers, Augsburg GmbH Germany	99.90%
49	4-n-Octylphenol	1806-26-4	Oc1ccc(CCCCCCCC)cc1	206.32	Aldrich	99%
50	Triclosan (5-Chlor-2-(2,4-dichlorphenoxy)-phenol)	3380-34-5	Oc2c(Oc1c(Cl)cc(Cl)cc1)cc(Cl)c2	289.55	Merck KGaA	98%
	Heterocyclic compounds					
51	Acridine	260-94-6	c3cc2cc1ccccc1nc2cc3	179.22	Merck KGaA	≥ 98 % GC
52	Pyridine	110-86-1	c1ccccn1	79.10	Merck KGaA	≥ 99.5 % GC
53	2-Methylpyridine	109-06-8	Cc1ccccn1	93.13	Merck KGaA	≥ 98 % GC
54	6-Nitroquinoline	613-50-3	[O-][N+](c2cc1cccnc1cc2)=O	174.16	Alfa Aesar GmbH & Co KG	98%
55	Ametryn	834-12-8	n1c(SC)nc(nc1NC(C)C)NCC	227.33	Riedel-de-Haën AG	98.2 % HPLC
56	Atrazine	1912-24-9	n1c(Cl)nc(nc1N(C)C)NCC	215.69	Riedel-de-Haën AG	97.4 % HPLC
57	Prometryn	7287-19-6	n1c(NC(C)C)nc(NC(C)C)nc1SC	241.36	Riedel-de-Haën AG	99.7 % GC
58	Propazine	139-40-2	n1c(Cl)nc(nc1NC(C)C)NC(C)C	229.71	Riedel-de-Haën AG	99.5 % HPLC
59	Sebuthylazine	7286-69-3	n1c(nc(nc1NCC)NC(C)C)Cl	229.71	Riedel-de-Haën AG	99.5 %
60	Simetryn	1014-70-6	CCNc1nc(SC)nc(NCC)n1	213.31	Riedel-de-Haën AG	99.2 %
61	Terbuthylazine	5915-41-3	n1c(Cl)nc(NC(C)(C)C)nc1NCC	229.71	Riedel-de-Haën AG	99.2 %
62	Terbutryn	886-50-0	n1c(nc(nc1NC(C)(C)C)SC)NCC	241.36	Riedel-de-Haën AG	98.7 % HPLC

9.3 List of Buffers and Solvents

No.	Name	CAS No.	M.Wt. g.mol⁻¹	Provided Company	Purity
1	2-Propanol	67-63-0	60.0956	Merck KGaA	99.70% GR for analysis
2	Acetonitrile	75-05-8	41.0524	Merck KGaA	99.90% GR for analysis
3	Cyclohexane	110-82-7	84.1608	Merck KGaA	99.50% GR for analysis
4	Di-Sodium hydrogen phosphate, Anhydrous	7558-79-4	141.96	Merck KGaA	99% GR for analysis
5	Hydrochloric acid (1N)	7647-01-0	36.4609	Merck KGaA	32% GR for analysis
6	Methanol	67-56-1	32.04	Merck KGaA	≥ 99.9 % GC GR for analysis
7	Potassium hydrogen phthalate	877-24-7	204.2151	Merck KGaA	99.90% GR for analysis
8	Sodium azide	26628-22-8	65.01	Merck KGaA	> 99 %
9	Sodium hydroxide solution (2N)	1310-73-2	39.99707	Merck KGaA	32% GR for analysis

9.4 List of Equipments

No.	Name	Provided Company
1-a	GCMS- System GC,MS5890 series gas chromatograph 5971 series mass selective detector Autosampler CP-8200 Capillary column is CP-Sil 8CB-MS 50m length x 0,32 mm ID x 0,12μm film thickness (5% diphenyl + 95% dimethylsiloxane)	Hewlett Packard Hewlett Packard Varian GmbH, Germany Chrompack, Varian GmbH, Germany
1-b	GC₂MS 6890N seriesgas chromatograph 5973N series mass selective detector Multipurpose Autosampler Capillary column is HP-5MS 30m length x 0,25mm ID x 0, 25μm film thickness (5% phenyl + 95% dimethylsiloxane)	Agilent Technol, Germany Agilent Technol, Germany Gerstel, Germany Agilent Technol, Germany
1-c	GC₃MS 7890A series gas chromatograph 5975C series mass selective detector Multipurpose Autosampler Capillary column is HP-5MS 30m length x 0,25mm ID x 0, 25μm film thickness (5% phenyl + 95% dimethylsiloxane)	Agilent Technol, Germany Agilent Technol, Germany Gerstel, Germany Agilent Technol, Germany
2-a	Balance (1) AG245	Mettler-Toledo, Switzerland
2-b	Balance (2) Chyo JL-200	Chyo Balance, Japan
3	Schott pH-meter type CG842/14PH	SI Analytics GmbH, Germany
4	Elma Ultrasonic type T570/H	Elma, Germany
5	Stirrer Heidolph Stirrer type MR3000D	Heidolph Instruments GmbH & Co. KG, Germany
6	Heidolph Shaker type Promax 3020	Heidolph Instruments GmbH & Co. KG, Germany
7	Heraeus Incubator type BK 6160	Thermo Scientific, Germany
8	WTW Microprocessor conductivity meter LF-196	Wissenschaftliche- Technische, Wellheim, Germany
9	Carbon analysis by RC-412 Hydrogen analysis by Tru Spec CHN Sulphur analysis by SC- 444	LECO Instruments GmbH, Germany
10	System 2000 FTIR Spektrometer	Perkin Elmer, Massachusetts, USA

9.5 List of Materials

No.	Name	Provided Company
1-a	Vials (1) SPME (10, 15mL)	Varian GmbH, Germany
1-b	Vials (2) white &brawn (10, 20mL) with screw neck	Gerstel, Germany
2	Syringe (0.5 – 500 mL)	Agilent Technol, Germany
	SPME fibreassembly	
3-a	PDMS 100 μm with 23 gauge for manual holder	
3-b	PDMS 100 μm with 23 gauge for automated holder	
3-c	PA 65 μm with 23 gauge for manual holder	Supelco (Bellefonte, PA, USA).
3-d	PA 65 μm with 23 gauge for automated holder	
3-e	PDMS-DVB 65μm with 23 gauge for automated holder	
4	Silicon blue transport septum, 17.5mm for (10, 20mL)	Gerstel, Germany
5	Sealing disks N20 septum red transparent	Macherey-Nagel GmbH & Co.KG, Germany
6	Open centre seals 20mm for (5-100mL vials)	Supelco, USA
7	Screw cap 18mm with septum	Gerstel, Germany

9.6 List of Computer Software

No.	Name	Provider Information
1	Microsoft office v.2002 (10.6854.6845)	Microsoft Corporation 1983-2001
2	ChemDraw Ultra v.10.0	CambridgeSoft 1986-2005, United Kingdom
3	GC₂MS Software (HP 5972/5971)MSD Chemstation G1034C v.C.03.00	Hewlett-Packard 1989-1994
4	GC₂MS Software (Enhanced Chemstation) MSD Chemstation D.03.00.611	Agilent Technologies 1989-2006
5	GC₃MS Software (Enhanced Chemstation) MSD Chemstation E.02.00.493	Agilent Technologies 1989-2008
6	Microcal™ Origin version 6.0	Microcal Software, Inc. 1991-1999, Northampton, USA
7	OriginPro 8G SR2 v.8.0891 (B 891)	OriginLab Corporation 1991-2008, Northampton, USA
8	STATISTICA v.8.0	StatSoft, Inc. 1984-2008, Tulsa, USA
9	ChemProp v.5.2.5 http://www.ufz.de/index.php?en=6738	"UFZ in-house version of ChemProp"
10	R Console version 2.10.1 (2009-12-14)	The R Foundation for Statistical Computing 2009
11	NIST 98 databank	NIST MS Library, Scientific Instrument Services, Inc., USA
12	ACD/pK$_a$ v.12	Advanced Chemistry Development (Company) 2009, Toronto, Canada.
13	Epi suite ™ v.3.20 includes PCKOCWIN v.1.66	U.S Environmental Protection Agency 2000-2007
14	Epi suite ™v.4	U.S Environmental Protection Agency 2008
15	Chemfinder database http://chembiofinder.cambridgesoft.com/chembiofinder/SimpleSearch.aspx	CambridgeSoft Corporation
16	Environmental science-Interactive physprop database http://www.syrres.com/what-we-do/databaseforms.aspx?id=386	Syracuse Research Corporation (SRC)
17	Online converting Henry's law constants http://dionysos.mpch-mainz.mpg.de/~sander/res/henry-conv.html January 2008	**Rolf Sander**, Air Chemistry Department, Max-Planck Institute for Chemistry, Mainz, Germany

9.7 The Physico-chemical Properties of Probe Compounds

No.	Substances	CAS No.	log S_w (exp.)	log K_{ow} (exp.)	pK_a (exp.)	ChemProp. log K_{aw} (est.)	Henry's law Constant unitless (volatility) [dim - less gas/ aqu] at 25 °C
1	1,2,3,4-Tetrachlorobenzene	634-66-2	0.77	4.60	n.a.	-1.18	$6.62 \cdot 10^{-2}$
2	1,2,4-Trimethylbenzene	95-63-6	1.76	3.63	n.a.	0.22	$2.96 \cdot 10^{-1}$
3	1,3,5-Trichlorobenzene	108-70-3	1.48	4.19	n.a.	-1.04	$7.48 \cdot 10^{-2}$
4	1-Aminopyrene	1606-67-3	-0.24#	4.31	4.32#	-6.92	$1.15 \cdot 10^{-7}$
5	1-Nitronaphthalene	86-57-7	0.96	3.19	n.a.	-4.07	$8.50 \cdot 10^{-5}$
6	2,3,4,6-Tetrachlorophenol	58-90-2	1.36	4.45	5.22	-0.83	$3.50 \cdot 10^{-4}$
7	2,3,5,6-Tetrachloraniline	3481-20-7	0.98#	4.10	-1.4#	-5.16	$2.27 \cdot 10^{-5}$
8	2,3-Dimethylnaphthalene	581-40-8	0.30	4.40	n.a.	-4.63	$3.64 \cdot 10^{-2}$
9	2,4,6-Trichlorophenol	88-06-2	2.90	3.69	6.23	-4.90	$1.03 \cdot 10^{-4}$
10	2,4-Dichlorophenol	120-83-2	3.65	3.06	7.89	-5.03	$1.70 \cdot 10^{-4}$
11	2,4-Di-t-butylphenol	96-76-4	1.54	5.19	11.7	-3.06	$1.48 \cdot 10^{-4}$
12	2-Aminoanthracene	613-13-8	0.11	3.43#	4.32#	-6.13	$2.30 \cdot 10^{-10}$
13	2-Fluorobiphenyl	321-60-8	1.40#	3.96#	n.a.	-1.70	$1.97 \cdot 10^{-2}$
14	2-Methylanthracene	613-12-7	-1.67	5.00	n.a.	-1.88	$2.32 \cdot 10^{-3}$
15	2-Methylpyridine	109-06-8	6.00	1.11	6	-2.75	$3.94 \cdot 10^{-4}$
16	2-Nitrofluorene	607-57-8	-0.67	3.37	n.a.	-4.29	$2.70 \cdot 10^{-5}$
17	3-Phenylphenol	580-51-8	2.61#	3.23	9.64	-5.75	$1.70 \cdot 10^{-6}$
18	4,4'-Isopropylidenediphenol	80-05-7	2.08	3.32	10.29#/10.93#	-8.68	$3.96 \cdot 10^{-10}$
19	4-Chloroaniline	106-47-8	3.59	1.83	3.98	-4.24	$4.59 \cdot 10^{-5}$
20	4-Chlorophenol	106-48-9	4.38	2.39	9.41	-4.77	$1.70 \cdot 10^{-5}$
21	4-Hexylphenol	2446-69-7	1.47#	4.52#	n.a.	-3.23	$8.95 \cdot 10^{-2}$
22	4-Nitrophenol	100-02-7	4.06	1.91	7.15	-7.04	$1.64 \cdot 10^{-8}$
23	4-n-Nonylphenol	104-40-5	0.80	5.76	10.15#	-2.86	$1.35 \cdot 10^{-3}$
24	4-Octylphenol	1806-26-4	0.49#	5.50#	n.a.	-2.98	$1.84 \cdot 10^{-4}$
25	6-Aminochrysene	2642-98-0	-0.81	4.99	4.32#	-7.14	$7.01 \cdot 10^{-8}$
26	6-Nitroquinoline	613-50-3	2.82#	1.84	3.24#	-6.95	$1.11 \cdot 10^{-7}$
27	Acridine	260-94-6	1.58	3.40	5.45	-5.56	$2.75 \cdot 10^{-6}$
28	Aldrin	309-00-2	-1.77	6.50	n.a.	-0.92	$1.74 \cdot 10^{-3}$
29	Ametryne	834-12-8	2.32	2.98	4.1	-10.94	$9.62 \cdot 10^{-8}$
30	Aniline	62-53-3	4.56	0.90	4.6	-4.11	$7.99 \cdot 10^{-5}$
31	Anthracene	120-12-7	-1.36	4.45	n.a.	-2.68	$4.71 \cdot 10^{-11}$
32	Atrazine	1912-24-9	1.54	2.61	1.7	-9.53	$9.34 \cdot 10^{-8}$

Chapter 9 - Appendix

33	Benzaldehyde	100-52-7	3.84	1.48	14.9	-2.31	5.48·10⁻⁴
34	Benzene	71-43-2	3.25	2.13	n.a.	-0.66	2.20·10⁻¹
35	Benzo(a)pyrene	50-32-8	-2.79	6.13	n.a.	-4.48	1.81·10⁻⁵
36	Fluoranthene	206-44-0	-0.59	5.16	n.a.	-3.47	3.51·10⁻⁴
37	Fluorene	86-73-7	0.23	4.18	n.a.	-1.88	6.83·10⁻³
38	Fluorobenzene	462-06-6	3.19	2.27	n.a.	-0.59	2.57·10⁻¹
39	Hexachlorobenzene	118-74-1	-2.21	5.73	n.a.	-0.85	6.73·10⁻²
40	Lindane	58-89-9	0.86	3.72	n.a.	-1.44	2.03·10⁻⁴
41	Naphthalene	91-20-3	1.49	3.30	n.a.	-1.67	1.74·10⁻²
42	2,4-DDD	53-19-0	-1.00	5.87#	n.a.	-1.15	1.77·10⁻³
43	2,4-DDE	3424-82-6	-0.85	6#	n.a.	-1.05	1.44·10⁻³
44	2,4-DDT	789-02-6	-1.07	6.79#	n.a.	-1.60	6.25·10⁻⁴
45	4,4-DDD	72-54-8	-1.05	6.02	n.a.	-1.15	2.61·10⁻⁴
46	4,4-DDE	72-55-9	-1.40	6.51	n.a.	-1.05	1.65·10⁻³
47	4,4-DDT	50-29-3	-2.26	6.91	n.a.	-1.60	3.29·10⁻⁴
48	PCB 180	35065-29-3	-2.41	8.27#	n.a.	-2.68	2.07·10⁻²
49	PCB 202	2136-99-4	-3.83	7.73	n.a.	-2.81	1.54·10⁻³
50	Pentachlorobenzene	608-93-5	-0.08	5.17	4.7	-1.31	4.90·10⁻²
51	Pentachlorophenol	87-86-5	1.15	5.12	4.7	-5.29	9.70·10⁻⁷
52	Phenol	108-95-2	4.92	1.46	9.99	-4.64	1.32·10⁻⁵
53	Prometryne	7287-19-6	1.52	3.51	4.05	-10.82	5.22·10⁻⁷
54	Propazine	139-40-2	0.93	2.93	1.7	-9.41	1.82·10⁻⁷
55	Pyridine	110-86-1	6.00	0.65	5.23	-3.54	2.88·10⁻⁴
56	Sebuthylazine	7286-69-3	1.66#	2.61*	2.5#	-9.41	2.35·10⁻⁷
57	Simetryn	1014-70-6	2.65	2.80	4	-11.06	2.11·10⁻⁷
58	Terbuthylazine	5915-41-3	0.93	3.21	2	-9.41	2.43·10⁻⁷
59	Terbutryn	886-50-0	1.40	3.74	4.03#	-10.82	4.55·10⁻⁷
60	Toluene	108-88-3	2.72	2.73	n.a.	0.136	2.43·10⁻¹
61	Triclosan	3380-34-5	1.00	4.76	7.8#	-6.69	1.98·10⁻⁷

- Log S_w exp. and log K_{ow} exp. from (Epi suite TM v.4), # is the estimated value.
- p$K_{a,exp}$ from (Environmental science-Interactive physprop database http://www.syrres.com/what-we-do/databaseforms.aspx?id=386), # is the estimated values from ACD/pK$_a$ v.12.
- Log $K_{aw(es)}$ from ChemProp v.5.2.5.
- Henry's law Constant unitless from (Environmental science-Interactive physprop database http://www.syrres.com/what-we-do/databaseforms.aspx?id=386)
  Online converting Henry's law constants http://dionysos.mpch-mainz.mpg.de/~sander/res/henry-conv.html.
- n.a. is not available data

9.8 FTIR and GC-MSD Tests for the Used Aldrich Humic Acid

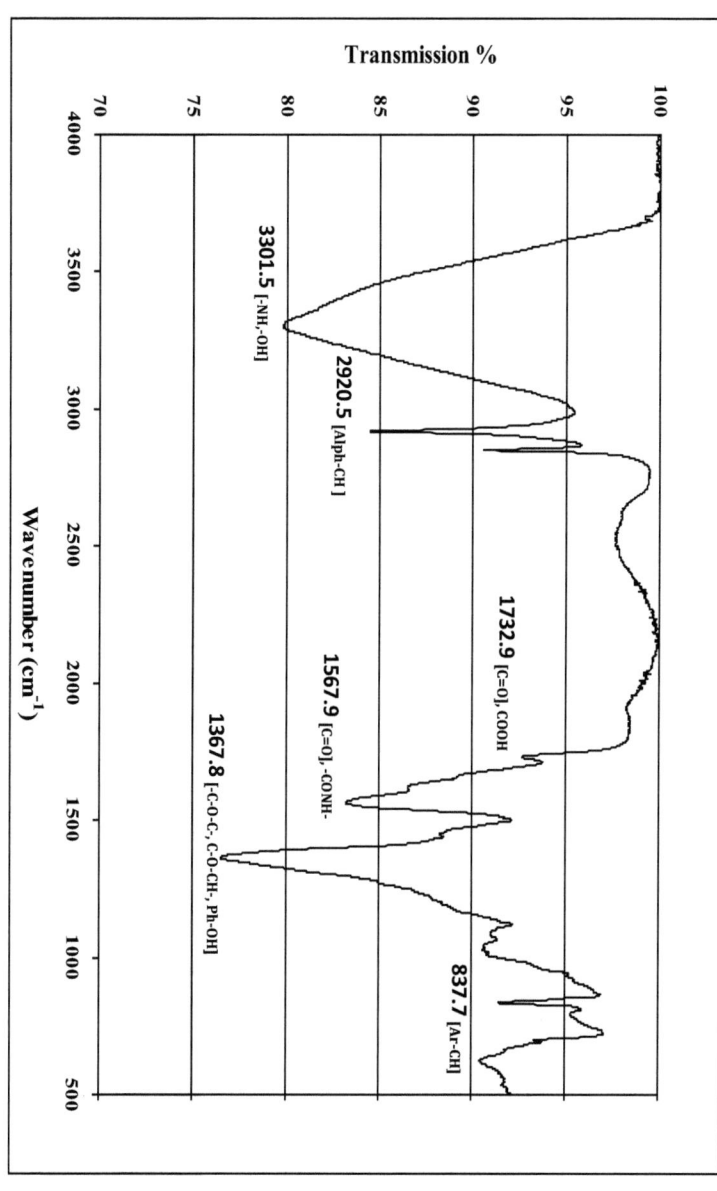

Figure 9.1 FTIR spectra of Aldrich humic acid.

Humic acid contains both hydrophilic and hydrophobic sites which are responsible for the reduction of solvent sorption, carboxylic and phenolic groups, 72% of the H-C groups of humic fraction exist as $(CH_2)_n$-CH_3 group. The long chain aliphatic structure present in HSs may contribute in the sorption of environmental solutes. HA has some region of polymeric three dimensional cross linked structure and its configuration can hinder the sorption process due to the restricted motion of crystalline or glassy state (site limitation of sorption can occur).

Choudhry G.G. *1984*. Humic substances, Structural, Photophysical, Photochemical and free radical, Aspects and interaction with environmental chemicals, University of Amsterdam. Network USA. *Cordon and Breach Science Publisher Inc.* pp: 100-131.

In this study, the FTIR spectrum of Aldrich HA as shown in Figure 9.1 was obtained by using Perkin Elmer FT-IR Spectrometer (system 2000). Figure 9.1 shows an FTIR spectrum of the Aldrich HA. The starching band obtained at 3301.5 and 2920.5 cm^{-1} may be attributed to the presence of amino groups or hydroxyl groups and aliphatic -CH- groups respectively. The band of carbonyl group appeared in 1732.9 and 1567.9 cm^{-1} due to the carboxyl and amide functional group in the HA. Another peak was also found at 1367.8 cm^{-1} and this may be attributed to the presence of either group -C-O-C-, -C-O-CH or the presence of phenolic groups. Aromatic -CH- peak appeared at 837.7 cm^{-1}.

Jonassess (2003) has demonstrate that the organic carbon content of Aldrich HA in the literature between 8-13% wt. to 69% wt. Infrared spectroscopy showed that the backbone of Aldrich HA consist of both aromatic and aliphatic part, and contains both carbonyl groups and carboxylic acids. Aldrich HA, is well characterized in the literature, consists of smaller molecular (approximately 4500 Da), with higher content of aromatic carbon. The presence of aromatic structure was detected by GC-MSD data and presence of some identified masses of PAHs such as specifically Naphthalene (m/z 128).

Jonassen K E N. *2003*. Determination of physico-chemical constants in sorption of polycyclic aromatic compounds to soil organic matter – investigated by use of soil organic matter HPLC columns. *PhD Thesis. Roskilde University, Denmark*.

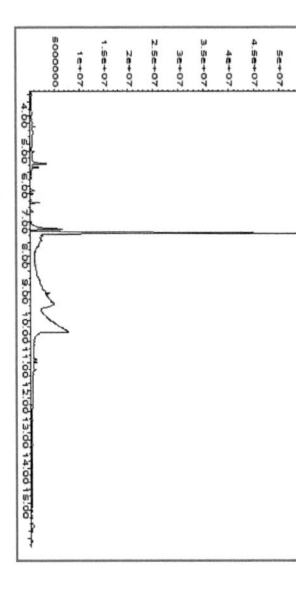

Figure 9.2 SCAN-GCMSD spectra of Aldrich humic acid.

Information of Aldrich humic acid

- The product was recovered from row lignite which contains a high amount of humic acid. Method is alkaline decomposition and isolation.

- Molecular weight 2000 - 500000

- Composition: this material consists of a mixture of complex macromolecules having polymeric phenolic structures. It is a colloidal acid product resulting from decomposition of organic matter obtained in open pit mining.

The GC-MSD data of different masses in Aldrich HA.

Peak No.	Ret. Time	Peak Area	Start Time	End Time	Name	CAS Number	Match Quality	Molecular Formula	M.Wt
1	4.306	7052931	4.28	4.334	Oxime-, methoxy-phenyl-	------------	83	$C_8H_9NO_2$	151.06
2	5.018	4187111	4.969	5.04	Ethanol, 2-(2-ethoxyethoxy)-	000111-90-0	86	$C_6H_{14}O_3$	134.09
3	5.356	89077185	5.243	5.406	Cyclotrisiloxane, hexamethyl-	000541-05-9	72	$C_6H_{18}O_3Si_3$	222.06
4	5.459	28878276	5.434	5.614	Benzene, 1-bromo-2-chloro-	000694-80-4	93	C_6H_4BrCl	189.92
5	6.125	22071713	6.026	6.175	Benzoic Acid	000065-85-0	94	$C_7H_6O_2$	122.04
6	6.207	7779998	6.175	6.259	Naphthalene	000091-20-3	95	$C_{10}H_8$	128.06
7	6.465	23955033	6.312	6.526	Cyclotetrasiloxane, octamethyl-	000556-67-2	47	$C_8H_{24}O_4Si_4$	296.07
8	6.716	17193295	6.545	6.785	n-Decanoic acid	000334-48-5	35	$C_{10}H_{20}O_2$	172.15
9	7.211	20592484	6.93	7.245	4-METHYL-1,2-BENZOQUINONE	004847-64-7	35	$C_7H_6O_2$	122.04
10	7.346	1532425881	7.245	7.371	4-Ethyl-3-methyl-9H-carbazole-2-carboxylate	071700-66-8	32	$C_{17}H_{17}NO_2$	267.13
11	7.377	26820495	7.371	7.401	2,3,5-Trioxabicyclo[2.1.0]pentane, 1,4-bis(chloromethyl)-	056247-52-0	43	$C_4H_4Cl_2O_3$	169.95
12	7.413	38571323	7.401	7.454	Phthalic anhydride	000085-44-9	97	$C_8H_4O_3$	148.02
13	7.46	10374242	7.454	7.478	Phthalic anhydride	000085-44-9	97	$C_8H_4O_3$	148.02
14	7.491	11928229	7.478	7.508	1,3-Isobenzofurandione	000085-44-9	50	$C_8H_4O_3$	148.02
15	7.539	18188494	7.508	7.584	1,3-Isobenzofurandione	000085-44-9	50	$C_8H_4O_3$	148.02
16	8.613	22780272	8.46	8.621	4-Pyridinecarbonitrile	000100-48-1	53	$C_6H_4N_2$	104.04
17	8.712	34857451	8.652	8.735	2H-Benzimidazol-2-one, 1,3-dihydro-5-methyl-	005400-75-9	38	$C_8H_8N_2O$	148.06
18	8.99	186764243	8.735	9.012	Thiourea	000062-56-6	64	CH_4N_2S	76.01
19	9.04	93423734	9.012	9.091	Benzaldehyde, 2,4-dihydroxy-3,6-dimethyl-	034883-14-2	83	$C_9H_{10}O_3$	166.06
20	9.358	599300772	9.091	9.54	Methane, chloro	000074-87-3	59	CH_3Cl	49.99
21	10.128	1108754290	9.54	10.173	1,2-Benzenedicarboxylic acid, monomethyl ester	004376-18-5	37	$C_9H_8O_4$	180.04
22	10.911	10243729	10.858	10.961	4-Isoxazolamine, 3-phenyl-	023350-02-9	43	$C_9H_8N_2O$	160.06
23	10.989	8984142	10.961	11.052	Berberine hydrochloride	000633-65-8	38	$C_{20}H_{18}ClNO_4$	371.09
24	11.212	9856615	11.132	11.281	9,10-Anthracenedione	000084-65-1	99	$C_{14}H_8O_2$	208.05

9.9 The Identification of Naphthalene in Humic Acid by GC-MSD

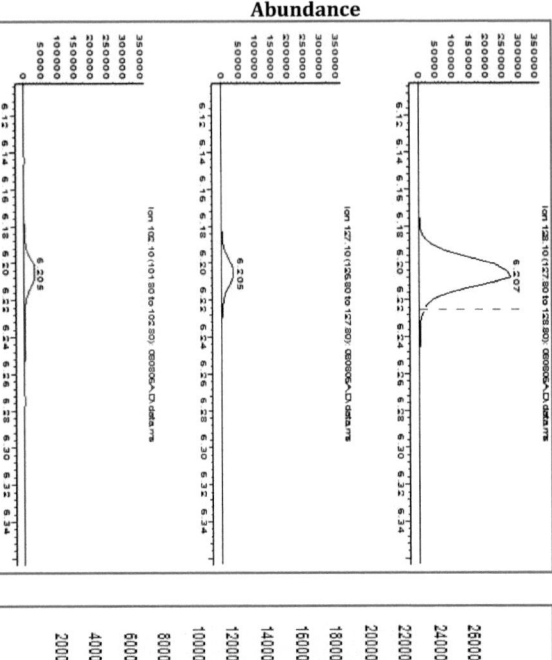

Figure 9.3 Naphthalene peak in Aldrich humic acid.

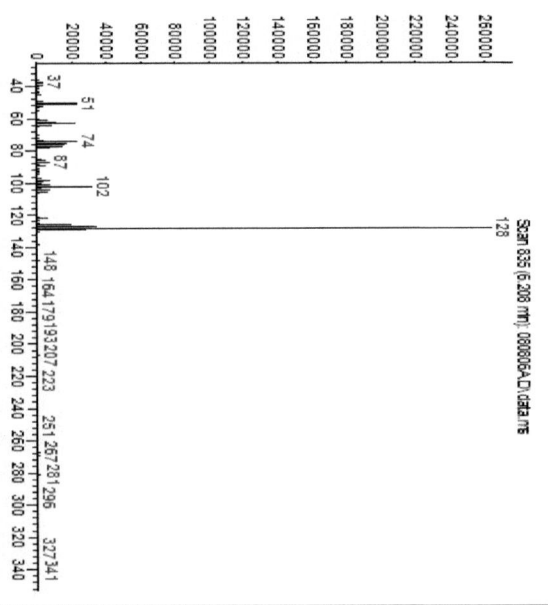

Figure 9.4 GC-MSD spectra of naphthalene.

9.10 The Sorption Experiments Conditions

No.	Substances	Equilibrium Time (h)	SPME Extraction Mode	Fibre Type	Extraction & Desorption Time (min)
1	1,2,3,4-Tetrachlorobenzene	24-36h	HS-mode	100μm PDMS	30 & 2
2	1,2,4-Trimethylbenzene	36-48h	HS-mode	100μm PDMS	10 & 2
3	1,3,5-Trichlorobenzene	24-36h	Direct mode	100 μm PDMS	30 & 2
4	1-Aminopyrene	24h	Direct mode	65 μm PDMS-DVB	30 & 2
5	1-Nitronaphthalene	24h	HS-mode	65μm PDMS-DVB	30 & 2
6	2,3,4,6-Tetrachlorophenol	24-36h	Direct mode	65 μm PA	30 & 2
7	2,3,5,6-Tetrachloraniline	24-36h	Direct mode	65 μm PDMS-DVB	30 & 2
8	2,3-Dimethylnaphthalene	24h	Direct mode	100 μm PDMS	60 & 2
9	2,4,6-Trichlorophenol	36-48h	Direct mode	65 μm PA	60 & 2
10	2,4-Dichlorophenol	36-48h	Direct mode	65 μm PA	30 & 2
11	2,4-Di-t-butylphenol	24-36h	Direct mode	65 μm PA	30 & 2
12	2-Aminoanthracene	24h	Direct mode	65 μm PDMS-DVB	20 & 3
13	2-Fluorobiphenyl	24h	HS-mode	100μm PDMS	30&2
14	2-Methylanthracene	24h	HS-mode	100μm PDMS	30 & 2
15	2-Methylpyridine	36-48h	Direct mode	65μm PDMS-DVB	30 & 3
16	2-Nitrofluorene	36-48h	HS-mode	65μm PDMS-DVB	30 & 2
17	3-Phenylphenol	36-48h	Direct mode	65 μm PA	20 & 3
18	4,4'-Isopropylidenediphenol	24-36h	Direct mode	65 μm PA	30 & 2
19	4-Chloroaniline	24-36h	Direct mode	65 μm PDMS-DVB	30 & 2
20	4-Chlorophenol	36-48h	HS-mode	65μm PA	20 & 2
21	4-n-Hexylphenol	36-48h	HS-mode	65μm PA	20 & 2
22	4-Nitrophenol	36-48h	Direct mode	65 μm PA	45 & 3
23	4-n-Nonylphenol	24-36h	Direct mode	65 μm PA	60 & 2

24	4-n-Octylphenol	36-48h	HS-mode	65μm PA	20 & 2
25	6-Aminochrysene	24h	Direct mode	65 μm PDMS-DVB	20 & 3
26	6-Nitroquinoline	36-48h	HS-mode	65μm PDMS-DVB	20 & 3
27	Acridine	24-36h	HS-mode	65μm PDMS-DVB	20 & 2
28	Aldrin	24h	Direct mode	100 μm PDMS	30 & 2
29	Ametryn	24-36h	Direct mode	65 μm PA	60 & 2
30	Aniline	36-48h	Direct mode	65 μm PDMS-DVB	30 & 2
31	Anthracene	24h	Direct mode	100 μm PDMS	60 & 2
32	Atrazine	24-36h	Direct mode	100 μm PDMS	60 & 2
33	Benzaldahyde	36-48h	HS-mode	65μm PDMS	20 & 3
34	Benzene	24-36h	Direct mode	100 μm PDMS	30 & 2
35	Benzo(a)pyrene	24h	Direct mode	100 μm PDMS	60 & 2
36	Fluoranthene	24h	Direct mode	100 μm PDMS	30 & 2
37	Fluorene	24-36h	HS-mode	100μm PDMS	20 & 3
38	Fluorobenzene	36-48h	HS-mode	100μm PDMS	10 & 3
39	Hexachlorobenzene	24h	Direct mode	100 μm PDMS	30 & 2
40	Lindane	24h	Direct mode	100 μm PDMS	30 & 2
41	Naphthalene	24h	Direct mode	100μm PDMS	30 & 2
42	2,4'-DDD	24h	HS-mode	100μm PDMS	30 & 2
43	2,4'-DDE	24h	HS-mode	100μm PDMS	30 & 2
44	2,4'-DDT	24h	HS-mode	100μm PDMS	30 & 2
45	4,4'-DDD	24h	Direct mode	100μm PDMS	30 & 2
46	4,4'-DDE	24h	Direct mode	100 μm PDMS	30 & 2
47	4,4'-DDT	24-36h	Direct mode	100 μm PDMS	30 & 2
48	PCB 180	24h	Direct mode	100μm PDMS	60 &3
49	PCB 202	24h	Direct mode	100μm PDMS	60 &3
50	Pentachlorobenzene	24-36h	HS-mode	100μm PDMS	30 & 2
51	Pentachlorophenol	24-36 h	Direct mode	65 μm PA	60 & 2
52	Phenol	36-48h	Direct mode	65 μm PA	30 & 3

53	Prometryn	24-36h	Direct mode	65 µm PA	60 & 2
54	Propazine	24-36h	Direct mode	65 µm PA	60 & 2
55	Pyridine	24-48h	HS-mode	65 µm PDMS-DVB	10 & 3
56	Sebuthylazine	24-36h	Direct mode	65 µm PA	60 & 2
57	Simetryn	24-48h	HS-mode	65 µm PA	30 & 2
58	Terbuthylazine	24-48h	HS-mode	65 µm PA	30 & 2
59	Terbutryn	24-36h	Direct mode	65 µm PA	60 & 2
60	Toluene	24-48h	HS-mode	100 µm PDMS	10 & 2
61	Triclosan	24-36h	Direct mode	65 µm PA	15 & 3

Log K_{oc} data of substances have been measured by direct mode; they were calculated from equation 3.8 in chapter 3. While of those have been measured by HS-Mode, they were calculated from equation 3.6 in chapter 3.

The used concentrations of humic acid

Chemical Classes	Conc. of humic acid
• Anilines	(1.5 : 10 gL⁻¹)
• Benzene & substituted benzenes (-CHO, -CH₃, 3(CH₃) and -F)	(0.6 : 20 gL⁻¹)
• Chlorinated Benzenes	(0.3 : 0.6 gL⁻¹)
• Biphenyls	(0.00025 : 0.6 gL⁻¹)
• Organochlorines	(0.015 : 0.3 gL⁻¹)
• PAHs & substituted PAHs	(0.02 : 1.5 gL⁻¹)
• Phenol & chlorophenols	(0.3 : 10 gL⁻¹)
• Alkylphenols	(0.05 : 0.6 gL⁻¹)
• Triazines	(0.6 : 1.8 gL⁻¹)

9.11 The Mass Spectroscopy Detections

No.	Substances	Stationary Phase	Retention Time (min)	Target Ion® (m/z)	Qualifiers® (m/z)	Temperature Program
1	1,2,3,4-Tetrachlorobenzene	Phase1	9.62	216	214, 218	(16)
2	1,2,4-Trimethylbenzene	Phase2	7.79	105	120, 77	(14)
3	1,3,5-Trichlorobenzene	Phase1	6.38	180	182, 145, 184	(3)
4	1-Aminopyrene	Phase2	16.46	217	189, 218, 216	(10)
5	1-Nitronaphthalene	Phase1	10.19	127	115, 173	(19)
6	2,3,4,6-Tetrachlorophenol	Phase2	9.25	232	230, 234, 131	(2)
7	2,3,5,6-Tetrachloroaniline	Phase2	9.38	231	229, 232	(7)
8	2,3-Dimethylnaphthalene	Phase2	8.12	156	141, 115, 128	(2)
9	2,4,6-Trichlorophenol	Phase2	11.17	196	198, 132, 160	(2)
10	2,4-Dichlorophenol	Phase2	6.01	162	164, 63, 98	(2)
11	2,4-Di-t-butylphenol	Phase1	9.54	191	192, 206	(3)
12	2-Aminoanthracene	Phase2	13.18	193	165, 194	(6)
13	2-Fluorobiphenyl	Phase1	8.42	172	171, 170	(14)
14	2-Methylanthracene	Phase1	12.84	192	191, 193, 189	(16)
15	2-Methylpyridine	Phase2	5.92	93	66, 78, 92	(8)
16	2-Nitrofluorene	Phase2	15.85	165	211, 164	(20)
17	3-Phenylphenol	Phase2	9.73	170	141, 115	(2)
18	4,4'-Isopropylidenediphenol	Phase1	14.11	213	228, 214	(3)
19	4-Chloroaniline	Phase2	6.62	127	129, 92, 65	(7)
20	4-Chlorophenol	Phase1	7.98	128	130, 65	(11)
21	4-n-Hexylphenol	Phase1	10.22	107	178, 77	(11)
22	4-Nitrophenol	Phase2	9.10	65	139, 109	(6)
23	4-n-Nonylphenol	Phase1	11.86	107	220, 77	(3)

24	4-n-Octylphenol	Phase1	11.75	107	108, 206	(11)
25	6-Aminochrysene	Phase2	23.48	243	215, 244	(9)
26	6-Nitroquinoline	Phase2	10.25	174	128, 116, 101	(17)
27	Acridine	Phase1	11.85	179	178, 151, 89	(11)
28	Aldrin	Phase1	12.46	263	293, 298	(4)
29	Ametryn	Phase1	15.14	227	212, 170	(3)
30	Aniline	Phase2	4.51	93	66, 65	(7)
31	Anthracene	Phase1	14.6	178	179, 176, 152	(3)
32	Atrazine	Phase1	11.04	200	215, 202, 173	(3)
33	Benzaldahyde	Phase2	4.94	106	105, 77	(17)
34	Benzene	Phase1	9.18	78	77, 51, 50	(1)
35	Benzo(a)pyrene	Phase1	27.19	252	253, 250, 126	(3)
36	Fluoranthene	Phase1	11.79	202	101, 203	(3)
37	Fluorene	Phase2	11.76	166	165, 82, 139	(14)
38	Fluorobenzene	Phase2	0.88	96	70, 75	(18)
39	Hexachlorobenzene	Phase1	10.90	284	286, 142	(3)
40	Lindane	Phase1	11.26	181	183, 219	(3)
41	Naphthalene	Phase2	6.87	128	129, 102	(2)
42	2,4'-DDD	Phase1	20.58	235	237, 165	(12)
43	2,4'-DDE	Phase1	19.75	246	248, 318, 316	(12)
44	2,4'-DDT	Phase1	21.34	235	237, 165	(12)
45	4,4'-DDD	Phase1	14.27	235	233, 165	(4)
46	4,4'-DDE	Phase1	13.71	246	318, 248, 316	(4)
47	4,4'-DDT	Phase1	14.64	235	237, 165	(4)
48	PCB 180	Phase2	14.20	394	396, 398, 324	(13)
49	PCB 202	Phase2	14.08	430	428, 432, 358	(13)
50	Pentachlorobenzene	Phase1	9.58	250	248, 252	(16)

51	Pentachlorophenol	Phase1	14.2	250	248, 252, 215	(3)
52	Phenol	Phase2	7.64	94	66, 65, 39	(5)
53	Prometryn	Phase1	12.07	241	184, 226	(3)
54	Propazine	Phase1	11.09	214	229, 172	(3)
55	Pyridine	Phase2	1.41	79	52, 51	(15)
56	Sebuthylazine	Phase1	14.6	200	202, 214	(3)
57	Simetryn	Phase1	12.14	213	170, 155	(11)
58	Terbuthylazine	Phase1	11.33	214	173, 132, 229	(11)
59	Terbutryn	Phase1	15.3	226	185, 170, 241	(3)
60	Toluene	Phase1	3.62	91	92, 65	(14)
61	Triclosan	Phase2	11.94	290	288, 218	(6)

Phase1: CP-Sil 8CB-MS (5% diphenyl + 95% dimethylsiloxane), Phase2: HP-5MS (5% phenyl + 95% dimethylsiloxane),@ is the data from NIST MS Library.

9.12 The Data of GC-MSD Temperature Programs

Temperature Program	Solvent Delay	Initial Temp. (°C)	Initial Time (min)	Ramp (1)			Ramp (2)		
				Rate (°C/min)	Final temp. (°C)	Final time (min)	Rate (°C/min)	Final temp. (°C)	Final time (min)
1	0	50°C	2	20	250	10	-----	-----	-----
2	3	70°C	2	20	250	20	-----	-----	-----
3	5	60°C	2	20	250	20	-----	-----	-----
4	0	70°C	2	20	250	30	-----	-----	-----
5	3	50°C	3	20	250	10	-----	-----	-----
6	5	70°C	2	20	250	15	-----	-----	-----
7	3	70°C	2	20	250	30	-----	-----	-----
8	3	30°C	2	10	180	5	20	250	0
9	5	70°C	2	20	250	25	-----	-----	-----
10	3	70°C	2	20	250	35	-----	-----	-----
11	5	70°C	2	20	250	10	-----	-----	-----
12	5	70°C	2	10	280	5	-----	-----	-----
13	15	140°C	2	2	170	0	15	320	1
14	2	70°C	2	10	310	10	-----	-----	-----
15	0	40°C	10	20	280	1	-----	-----	-----
16	5	70°C	2	10	250	10	-----	-----	-----
17	0.30	40°C	5	20	140	0	20	280	5
18	0.30	40°C	5	20	140	0	20	280	1
19	2	70°C	2	20	270	10	-----	-----	-----
20	3	70°C	3	10	250	10	-----	-----	-----

9.13 List of Experimental Linear Solvation Energy Relationship (LSER) Descriptors (Abraham Descriptors)

(4 decimal digits for modelling)

No.	Substances	CAS No.	E	S	A	BH	BO	L	V
1	1,2,3,4-Tetrachlorobenzene	634-66-2	1.1800	0.9200	0.0000	0.0000	0.0000	6.1710	1.2060
2	1,2,4-Trimethylbenzene	95-63-6	0.6800	0.5600	0.0000	0.1900	0.1900	4.4410	1.1391
3	1,3,5-Trichlorobenzene	108-70-3	0.9800	0.7300	0.0000	0.0000	0.0000	5.0450	1.0836
4	1-Aminopyrene	1606-67-3	n.a.	n.a.	n.a.	n.a.	n.a.	n.a.	1.6800
5	1-Nitronaphthalene	86-57-7	1.6000	1.5900	0.0000	0.2900	0.2900	6.9910	1.2596
6	2,3,4,6-Tetrachlorophenol	58-90-2	1.1000	0.8700	0.5000	0.1500	0.1500	6.7400	1.2647
7	2,3,5,6-Tetrachloroaniline	3481-20-7	1.3100	1.3400	0.4600	0.0300	0.0300	n.a.	1.3058
8	2,3-Dimethylnaphthalene	581-40-8	1.4000	0.8500	0.0000	0.2800	0.2800	6.3100	1.3672
9	2,4,6-Trichlorophenol	88-06-2	1.0100	1.0100	0.8200	0.0800	0.0800	5.6640	1.1423
10	2,4-Dichlorophenol	120-83-2	0.9600	0.9900	0.5800	0.1400	0.1400	4.9430	1.0199
11	2,4-Di-t-butylphenol	96-76-4	0.8300	0.9200	0.4400	0.5000	0.5000	n.a.	1.9000
12	2-Aminoanthracene	613-13-8	2.6400	2.0300	0.2100	0.3700	0.3700	9.3000	1.55
13	2-Fluorobiphenyl	321-60-8	n.a.	n.a.	n.a.	n.a.	n.a.	n.a.	1.3419
14	2-Methylanthracene	613-12-7	2.2900	n.a.	0.0000	0.3100	0.3100	8.1840	1.5953
15	2-Methylpyridine	109-06-8	0.6000	0.7500	0.0000	0.5800	0.4800	3.4200	0.8200
16	2-Nitrofluorene	607-57-8	1.8500	2.4400	0.0000	0.0000	0.0000	9.1260	1.5307
17	3-Phenylphenol	580-51-8	1.5600	1.4100	0.5900	0.4500	0.4500	n.a.	1.3829
18	4,4'-Isopropylidenediphenol	80-05-7	1.6100	1.5600	0.9900	0.9100	0.9100	9.6000	1.8600
19	4-Chloroaniline	106-47-8	1.0600	1.1300	0.3000	0.3100	0.3500	4.8900	0.9400
20	4-Chlorophenol	106-48-9	0.9200	1.0800	0.6700	0.2000	0.2000	4.7750	0.8975
21	4-n-Hexylphenol	2446-69-7	n.a.	n.a.	n.a.	n.a.	n.a.	n.a.	1.6205
22	4-Nitrophenol	100-02-7	1.0700	1.7200	0.8200	0.2600	0.2600	5.8800	0.9500
23	4-n-Nonylphenol	104-40-5	n.a.	n.a.	n.a.	n.a.	n.a.	n.a.	2.0432

#	Name	CAS								
24	4-n-Octylphenol	1806-26-4	0.7800	0.8800	0.5500	0.3700	n.a.	n.a.	n.a.	1.9923
25	6-Aminochrysene	2642-98-0	n.a.	n.a.	n.a.	n.a.	n.a.	n.a.	n.a.	1.9232
26	6-Nitroquinoline	613-50-3	n.a.	n.a.	n.a.	n.a.	n.a.	n.a.	n.a.	1.2185
27	Acridine	260-94-6	2.3600	1.3200	0.0000	0.5800	n.a.	n.a.	7.6440	1.4133
28	Aldrin	309-00-2	n.a.	n.a.	n.a.	n.a.	n.a.	n.a.	n.a.	2.0134
29	Ametryn	834-12-8	1.4700	1.2600	0.1700	1.0200	n.a.	1.0200	8.6070	1.8016
30	Aniline	62-53-3	0.9600	0.9600	0.2600	0.4100	0.5000	0.5000	3.9300	0.8200
31	Anthracene	120-12-7	2.2900	1.3400	0.0000	0.2800	0.2800	0.2800	7.5680	1.4544
32	Atrazine	1912-24-9	1.2200	1.2900	0.1700	1.0100	0.8800	0.8800	7.7830	1.6196
33	Benzaldahyde	100-52-7	0.8200	1.0000	0.0000	0.3900	0.3900	0.3900	4.0080	0.8730
34	Benzene	71-43-2	0.6100	0.5200	0.0000	0.1400	0.1400	0.1400	2.7860	0.7164
35	Benzo(a)pyrene	50-32-8	3.6300	1.9600	0.0000	0.3700	0.3700	0.3700	11.7360	1.9536
36	Fluoranthene	206-44-0	2.3800	1.5500	0.0000	0.2400	0.2400	0.2400	8.8270	1.5846
37	Fluorene	86-73-7	1.5900	1.0600	0.0000	0.2500	0.2500	0.2500	6.9220	1.3565
38	Fluorobenzene	462-06-6	0.4800	0.5700	0.0000	0.1000	0.1000	0.1000	2.7880	0.7341
39	Hexachlorobenzene	118-74-1	1.4900	0.9900	0.0000	0.0000	0.0000	0.0000	7.3900	1.4508
40	Lindane	58-89-9	1.4500	1.2800	0.0000	0.5000	0.5000	0.5000	7.5700	1.5798
41	Naphthalene	91-20-3	1.3400	0.9200	0.0000	0.2000	0.2000	0.2000	5.1610	1.0854
42	2,4'-DDD	53-19-0	n.a.	n.a.	n.a.	n.a.	n.a.	n.a.	n.a.	2.0956
43	2,4'-DDE	3424-82-6	n.a.	n.a.	n.a.	n.a.	n.a.	n.a.	n.a.	2.0526
44	2,4'-DDT	789-02-6	n.a.	n.a.	n.a.	n.a.	n.a.	n.a.	n.a.	2.2180
45	4,4'-DDD	72-54-8	n.a.	n.a.	n.a.	n.a.	n.a.	n.a.	n.a.	2.0956
46	4,4'-DDE	72-55-9	n.a.	n.a.	n.a.	n.a.	n.a.	n.a.	n.a.	2.0526
47	4,4'-DDT	50-29-3	1.8000	1.3800	0.0000	0.4000	0.4000	0.4000	9.5330	2.2180
48	PCB 180	35065-29-3	2.2900	1.8700	0.0000	0.0900	0.0900	0.0900	10.4150	2.1810
49	PCB 202	2136-99-4	2.4400	2.0000	0.0000	0.0600	0.0600	0.0600	10.1410	2.3034
50	Pentachlorobenzene	608-93-5	1.3300	0.9200	0.0600	0.0000	0.0000	0.0000	6.6300	1.3284

#	Name	CAS	E	S	A	B	V	L
51	Pentachlorophenol	87-86-5	1.2200	0.8600	0.6100	0.0900	0.0900	6.8220	1.3871
52	Phenol	108-95-2	0.8100	0.8900	0.6000	0.3000	0.3000	3.7700	0.7800
53	Prometryn	7287-19-6	1.4300	1.2300	0.1700	1.0100	1.0100	8.9640	1.9425
54	Propazine	139-40-2	1.1900	1.2600	0.1300	1.0500	0.9200	8.2630	1.7605
55	Pyridine	110-86-1	0.6300	0.8400	0.0000	0.5200	0.4700	3.0220	0.6753
56	Sebuthylazine	7286-69-3	n.a.	n.a.	n.a.	n.a.	n.a.	n.a.	1.7605
57	Simetryn	1014-70-6	1.5000	1.2900	0.2300	0.9900	0.9900	8.2310	1.6607
58	Terbuthylazine	5915-41-3	1.1900	1.2600	0.1400	0.9100	0.8500	8.0890	1.7605
59	Terbutryn	886-50-0	1.4300	1.2300	0.1200	0.9900	0.9900	8.9840	1.9425
60	Toluene	108-88-3	0.6000	0.5200	0.0000	0.1400	0.1400	3.3250	0.8573
61	Triclosan	3380-34-5	1.8500	1.6900	0.7400	0.2900	0.2900	n.a.	1.8100

E, S, A, B and V represent the compound descriptors for excess molar refraction (E) in cm³ mol⁻¹, dipolarity / polarisability (S), hydrogen bond acidity (A), hydrogen bond basicity (B) and McGowan volume of solute (V) in cm³ mol⁻¹. In case of a model that relies on processes involving a gas in condensed phase transfer depending on the solute but independent on the solvent, the logarithm of the solute gas phase dimensionless Ostwald partition coefficient in hexadecane at 289 K (L parameter) is used instead of V. V and L parameters are measured for the endoergic effect of disrupting solvent-solvent interactions. n.a. is not available.

Abraham MH, Chadha HS, Whiting GS, Mitchell RC. **1994a**. Hydrogen bonding. 32. An analysis of water-octanol and water-alkane partitioning and the Δ log parameter. of Seiler. *J Pharm. Sci.* 83:1085-1100.

Abraham MH, Haftvan JA, Whiting GS, Leo A, Taft RS. **1994b**. Hydrogen Bonding. Part 34. The factors that influence the solubility of gases and vapours in water at 298 K, and a new method for its determination. *J.Chem. Soc. Perkin Trans.* 2: 1777-1791.

Abraham MH, Ibrahim A, Zissimos AM. **2004**. Determination of sets of solute descriptors from chromatographic measurements. Review. *J of Chromatography A* 1037: 29-47.

Abraham MH, McGowan Jc. **1987**. The use of characteristic volumes to measure cavity terms in reversed phase liquid chromatography. *Chromatographia* 23:243-246.

Abraham MH, Poole C F, Poole SK.**1999**. Classification of stationary phases and other materials by gas chromatography. Review. *J. of Chromatography A* 842:79-114.

Abraham MH, Whiting GS, Doherty RM, Shuely WJ. **1991**. Hydrogen bonding. Part 13. A new method for the characterization of GLC stationary phases-the laffort data set. *J.Chem. Soc. Perkin Trans* 2:1451-1460.

Abraham MH. **1993**. Scales of solutes hydrogen-bonding: Their construction and application to physicochemical and biochemical processes. *Chemical Society Reviews*: 73-83.

Schüürmann G, Kühne R, Kleint F, Ebert RU, Rothenbacher C, Herth P. **1997**. "In-House software" ChemProp. Software system for automatic chemical property estimation from molecular structure. In: Chen F, Schüürmann G (eds), In Quantitative Structure-Activity Relationships in Environmental Sciences-VII. *SETAC Press: Pensacola*: 93-114.

9.14 List of Linear Solvation Energy Relationship (LSER) Descriptors (Platts Descriptors)

(4 decimal digits for modelling)

No.	Substances	CAS No.	E	S	A	BH	B0	L	V
1	1,2,3,4-Tetrachlorobenzene	634-66-2	1.1960	0.9040	0.0030	0.0700	0.0470	5.9450	1.2060
2	1,2,4-Trimethylbenzene	95-63-6	0.6800	0.5050	0.0030	0.1250	0.1000	4.3720	1.1391
3	1,3,5-Trichlorobenzene	108-70-3	1.0610	0.8620	0.0030	-0.0010	0.0010	5.3380	1.0836
4	1-Aminopyrene	1606-67-3	2.2830	1.8170	0.2500	0.4450	0.4980	9.6680	1.6844
5	1-Nitronaphthalene	86-57-7	1.3920	1.4390	0.0030	0.2910	0.2930	6.6220	1.2596
6	2,3,4,6-Tetrachlorophenol	58-90-2	1.2690	1.0490	0.9070	0.0890	0.0750	6.4010	1.2647
7	2,3,5,6-Tetrachloroaniline	3481-20-7	1.4710	1.2190	0.4100	0.2490	0.2700	6.9160	1.3058
8	2,3-Dimethylnaphthalene	581-40-8	1.1680	0.8310	0.0030	0.1510	0.1360	6.0820	1.3672
9	2,4,6-Trichlorophenol	88-06-2	1.1340	1.0400	0.7880	0.1190	0.1200	5.7650	1.1423
10	2,4-Dichlorophenol	120-83-2	0.9990	0.9830	0.7880	0.2000	0.1980	4.9930	1.0199
11	2,4-Di-t-butylphenol	96-76-4	0.8030	0.6690	0.5460	0.5270	0.5050	6.8930	1.9023
12	2-Aminoanthracene	613-13-8	1.9230	1.6150	0.2500	0.4450	0.4980	8.4200	1.5542
13	2-Fluorobiphenyl	321-60-8	1.1700	1.0300	0.0030	0.1700	0.1720	6.0760	1.3419
14	2-Methylanthracene	613-12-7	1.6560	1.1570	0.0030	0.1770	0.1720	7.7920	1.5953
15	2-Methylpyridine	109-06-8	0.6420	0.7260	0.0030	0.4430	0.4120	3.5250	0.8162
16	2-Nitrofluorene	607-57-8	1.7520	1.6410	0.0030	0.2910	0.2930	8.3690	1.5307
17	3-Phenylphenol	580-51-8	1.4610	1.2770	0.5460	0.4770	0.4830	6.8950	1.3829
18	4,4'-Isopropylidenediphenol	80-05-7	1.6130	1.4960	1.0890	0.8240	0.8290	8.8070	1.8643
19	4-Chloroaniline	106-47-8	1.0660	1.1440	0.4180	0.3440	0.3880	4.8720	0.9386
20	4-Chlorophenol	106-48-9	0.9640	1.0080	0.8980	0.3760	0.3730	4.5950	0.8975
21	4-n-Hexylphenol	2446-69-7	0.8370	0.8510	0.5460	0.4290	0.4230	6.7420	1.6205
22	4-Nitrophenol	100-02-7	1.0690	1.4350	0.8980	0.5190	0.5180	5.2630	0.9493

#	Name	CAS							
23	4-n-Nonylphenol	104-40-5	0.8370	0.8510	0.5460	0.4290	0.4230	8.2390	2.0432
24	4-n-Octylphenol	1806-26-4	0.8370	0.8510	0.5460	0.4290	0.4230	7.7400	1.9023
25	6-Aminochrysene	2642-98-0	2.4190	1.9170	0.2500	0.4670	0.5220	10.6060	1.9232
26	6-Nitroquinoline	613-50-3	1.3700	1.6120	0.0030	0.6010	0.5810	6.7270	1.2185
27	Acridine	260-94-6	1.6260	1.3540	0.0030	0.4910	0.4720	7.4210	1.4133
28	Aldrin	309-00-2	2.1150	2.4070	0.0030	0.5530	0.5510	10.5470	2.0134
29	Ametryn	834-12-8	2.0530	2.0680	0.3910	1.6090	1.4440	9.7600	1.8016
30	Aniline	62-53-3	0.9310	1.0110	0.2500	0.4010	0.4500	4.0480	0.8162
31	Anthracene	120-12-7	1.6480	1.1810	0.0030	0.1810	0.1840	7.3160	1.4544
32	Atrazine	1912-24-9	1.8500	2.0360	0.3910	1.4340	1.2690	8.8500	1.6196
33	Benzaldahyde	100-52-7	0.7950	1.0480	0.0030	0.4710	0.4750	4.0630	0.8730
34	Benzene	71-43-2	0.6560	0.5770	0.0030	0.1370	0.1360	2.9440	0.7164
35	Benzo(a)pyrene	50-32-8	2.5040	1.6850	0.0030	0.2030	0.2080	10.7500	1.9536
36	Fluoranthene	206-44-0	2.0080	1.3830	0.0030	0.1810	0.1840	8.5640	1.5846
37	Fluorene	86-73-7	1.5120	1.0810	0.0030	0.1590	0.1600	6.8770	1.3565
38	Fluorobenzene	462-06-6	0.5380	0.6280	0.0030	0.1260	0.1240	2.9520	0.7341
39	Hexachlorobenzene	118-74-1	1.4660	0.8890	0.0030	-0.0910	-0.1340	7.2460	1.4508
40	Lindane	58-89-9	0.9920	0.9850	0.1170	0.2150	0.2440	7.1620	1.5798
41	Naphthalene	91-20-3	1.1520	0.8790	0.0030	0.1590	0.1600	5.1300	1.0854
42	2,4'-DDD	53-19-0	1.7650	1.4310	0.0530	0.1670	0.1840	9.8410	2.0956
43	2,4'-DDE	3424-82-6	1.9640	1.6110	0.0030	0.1590	0.1600	10.3020	2.0526
44	2,4'-DDT	789-02-6	1.9030	1.5980	0.0030	0.2070	0.2270	10.6150	2.2180
45	2,4'-DDD	72-54-8	1.7650	1.4310	0.0530	0.1670	0.1840	9.8410	2.0956
46	2,4'-DDE	72-55-9	1.9640	1.6110	0.0030	0.1590	0.1600	10.3020	2.0526
47	2,4'-DDT	50-29-3	1.9030	1.5980	0.0030	0.2070	0.2270	10.6150	2.2180
48	PCB 180	35065-29-3	2.2330	1.6240	0.0030	0.0310	0.0010	11.4340	2.1800
49	PCB 202	2136-99-4	2.3680	1.7190	0.0030	-0.1070	-0.1440	12.2320	2.3034

#	Name	CAS	E	S	A	B	V	L
50	Pentachlorobenzene	608-93-5	1.3310	0.9180	0.0030	-0.0260	-0.0560	6.6360	1.3284
51	Pentachlorophenol	87-86-5	1.4040	1.0200	1.0260	0.0240	-0.0030	7.0110	1.3871
52	Phenol	108-95-2	0.8300	0.8800	0.5500	0.4300	0.4400	3.7700	0.7800
53	Prometryn	7287-19-6	2.0380	2.0290	0.3910	1.6270	1.4640	10.0310	1.9425
54	Propazine	139-40-2	1.8350	1.9970	0.3910	1.4520	1.2890	9.1210	1.7605
55	Pyridine	110-86-1	0.6340	0.7500	0.0030	0.4470	0.4240	3.0490	0.6753
56	Sebuthylazine	7286-69-3	1.8500	2.0360	0.3910	1.4340	1.2690	9.3490	1.7605
57	Simetryn	1014-70-6	2.0680	2.1070	0.3910	1.5910	1.4240	9.4890	1.6607
58	Terbuthylazine	5915-41-3	1.8440	1.9960	0.3910	1.4670	1.2960	9.1650	1.7605
59	Terbutryn	886-50-0	2.0470	2.0280	0.3910	1.6420	1.4710	10.0750	1.9425
60	Toluene	108-88-3	0.6640	0.5530	0.0030	0.1330	0.1240	3.4200	0.8573
61	Triclosan	3380-34-5	1.8110	1.4620	0.4140	0.4760	0.4350	9.5130	1.8088

E, S, A, B and V represent the compound descriptors for excess molar refraction (E) in cm^3 mol^{-1}, dipolarity / polarisability (S), hydrogen bond acidity (A), hydrogen bond basicity (B) and McGowan volume of solute (V) in cm^3 mol^{-1}. In case of a model that relies on processes involving a gas in condensed phase transfer depending on the solute but independent on the solvent, the logarithm of the solute gas phase dimensionless Ostwald partition coefficient in hexadecane at 289 K (L parameter) is used instead of V. V and L parameters are measured for the endoergic effect of disrupting solvent-solvent interactions. n.a. is not available.

- Platts JA, Butina D, Abraham MH, Hersey A. **1999**.Estimation of molecular linear free energy relationship descriptors using a group contribution approach. *J. Chem. Inf Comput. Sci.* 39(5): 835-845.

- Schüürmann G, Kühne R, Kleint F, Ebert RU, Rothenbacher C, Herth P. **1997**."In-House software" ChemProp. Software system for automatic chemical property estimation from molecular structure.In: Chen F, Schüürmann G (eds), In Quantitative Structure-Activity Relationships in Environmental Sciences-VII. SETAC Press: Pensacola: 93-114.

9.15 Cross-Correlation of LSER Parameters

LSER exp. input parameters

	E1	S1	A1	B01	V1
E1	1	0.611	0.042	0.001	0.392
S1	0.611	1	0.003	0.012	0.346
A1	0.042	0.003	1	0.004	0.001
B01	0.001	0.012	0.004	1	0.136
V1	0.392	0.346	0.001	0.136	1

	E1	S1	A1	B01	V1
E1	1	0.61	0.046	0.002	0.318
S1	0.61	1	0.033	0.065	0.295
A1	0.046	0.033	1	0.004	0.002
B01	0.002	0.065	0.004	1	0.33
V1	0.318	0.295	0.002	0.33	1

	E1	S1	A1	B01	V1
E1	1	0.71	0.085	0.009	0.57
S1	0.71	1	0.027	0.032	0.49
A1	0.085	0.027	1	0.049	0.00
B01	0.009	0.032	0.049	1	0.00
V1	0.57	0.49	0.00	0.00	1

LSER calc. input parameters

Sorption in neutral water
(25±°C, pH=7±0.2)

	E2	S2	A2	B02	V2
E2	1	0.754	0.017	0.089	0.569
S2	0.754	1	0.00	0.405	0.403
A2	0.017	0.00	1	0.095	0.003
B02	0.089	0.405	0.095	1	0.045
V2	0.569	0.403	0.003	0.045	1

pH-dependence of sorption
(25±°C, pH=7±0.2)

	E2	S2	A2	B02	V2
E2	1	0.739	0.0418	0.148	0.513
S2	0.739	1	0.001	0.487	0.40
A2	0.0418	0.001	1	0.036	0.017
B02	0.148	0.487	0.036	1	0.12
V2	0.513	0.40	0.017	0.12	1

Temperature-dependence
of sorption
(25±°C, pH=7±0.2)

	E2	S2	A2	B02	V2
E2	1	0.77	0.026	0.01	0.65
S2	0.77	1	0.00	0.24	0.415
A2	0.026	0.00	1	0.214	0.00
B02	0.012	0.24	0.214	1	0.00
V2	0.65	0.415	0.00	0.00	1

9.16 The Experimental Values of log K_{oc} Relevant to Humic acids, Soils and Sediments from Literature

No.	Substances	CAS No.	Exp. log K_{oc} Aldrich humic acid in the current study	Exp. log K_{oc} (not specified) in literature			References
				Range, n	Mean value of Exp. log K_{oc}, ±SD	Min - Max	All
1	1,2,3,4-Tetrachlorobenzene	634-66-2	4.50	(3.15,...,4.53), 10	3.92, ±0.38	(k) - (k)	(w), (b), (e), (f), (h)², (k)³, (v)
2	1,2,4-Trimethylbenzene	95-63-6	3.38	1	3.60	n.a.- n.a.	(b)
3	1,3,5-Trichlorobenzene	108-70-3	3.02	(2.85,...,4.46), 8	3.48, ±0.69	(w) - (k)	(w), (c), (f), (h)², (k)³
4	1-Aminopyrene	1606-67-3	4.82	n.a.	n.a.	n.a.- n.a.	n.a.
5	1-Nitronaphthalene	86-57-7	3.00	n.a.	n.a.	n.a.- n.a.	n.a.
6	2,3,4,6-Tetrachlorophenol	58-90-2	2.90	(2.35,...,3.79), 9	3.26, ±0.50	(k) - (f)	(w), (c), (e), (f), (h)², (k)³
7	2,3,5,6-Tetrachloraniline	3481-20-7	3.81	1	3.94	n.a.- n.a.	(c)
8	2,3-Dimethylnaphthalene	581-40-8	3.68	n.a.	n.a.	n.a.- n.a.	n.a.
9	2,4,6-Trichlorophenol	88-06-2	2.37	(1.97,...,3.07), 11	2.83, ±0.35	(k) - (k)	(w), (b), (c), (e), (f), (h)², (k)³, (v)
10	2,4'-DDD	53-19-0	5.57	n.a.	n.a.	n.a.- n.a.	n.a.
11	2,4'-DDE	3424-82-6	5.41	1	5.49	n.a.- n.a.	(*)
12	2,4'-DDT	789-02-6	5.13	n.a.	n.a.	n.a.- n.a.	n.a.
13	2,4-Dichlorophenol	120-83-2	2.60	(1.93,...,2.81)	2.59, ±0.27	(k) - (w)	(w), (c), (e), (f), (h)², (k)³, (q), (v)
14	2,4-Di-t-butylphenol	96-76-4	3.99	n.a.	n.a.	n.a.- n.a.	n.a.
15	2-Aminoanthracene	613-13-8	4.75	1	4.45	n.a.- n.a.	(c)
16	2-Fluorobiphenyl	321-60-8	3.81	n.a.	n.a.	n.a.- n.a.	n.a.

#	Compound	CAS					
17	2-Methylanthracene	613-12-7	4.50	n.a.	n.a.	n.a.-n.a.	n.a.
18	2-Methylpyridine	109-06-8	2.88	n.a.	n.a.	n.a.-n.a.	n.a.
19	2-Nitrofluorene	607-57-8	3.42	n.a.	n.a.	n.a.-n.a.	n.a.
20	3-Phenylphenol	580-51-8	3.33	n.a.	n.a.	n.a.-n.a.	n.a.
21	4,4'-DDD	72-54-8	5.02	(4.21....5.38), 4	4.63, ±0.55	(z) - (b)	(z), (w), (b), (e)
22	4,4'-DDE	72-55-9	5.02	(4.70....4.82), 4	4.79, ±0.06	(b) - (w)	(w), (b), (c), (e)
23	4,4'-DDT	50-29-3	5.45	(5.28....5.68), 8	5.41, ±0.16	(k) - (k)	(w), (b), (c), (e), (h), (k)³
24	4,4'-Isopropylidenediphenol	80-05-7	3.44	n.a.	n.a.	n.a.-n.a.	n.a.
25	4-Chloroaniline	106-47-8	3.61	(1.96...1.98), 4	1.97, ±0.01	(w) - (b)	(w), (b), (c), (e)
26	4-Chlorophenol	106-48-9	2.96	1	1.85	n.a.-n.a.	(b)
27	4-n-Hexylphenol	2446-69-7	3.98	n.a.	n.a.	n.a.-n.a.	n.a.
28	4-Nitrophenol	100-02-7	2.62	(1.74...2.37), 8	2.16, ±0.27	(b) - (w)	(w), (b), (c), (e), (k)³, (q)
29	4-n-Nonylphenol	104-40-5	4.87	1	3.84	n.a.-n.a.	(z)
30	4-n-Octylphenol	1806-26-4	3.98	n.a.	n.a.	n.a.-n.a.	n.a.
31	6-Aminochrysene	2642-98-0	5.34	(5.16...5.21), 6	5.19, ±0.03	(b) - (*)	(*), (b), (c), (f), (h)²
32	6-Nitroquinoline	613-50-3	3.25	n.a.	n.a.	n.a.-n.a.	n.a.
33	Acridine	260-94-6	3.63	(4.11...4.18), 6	4.14, ±0.03	(b) - (w)	(w), (b), (e), (f), (h), (v)
34	Aldrin	309-00-2	5.23	(4.10...4.69), 4	4.54, ±0.29	(b) - (w)	(w), (b), (c), (e)
35	Ametryne	834-12-8	2.70	(2.27....2.83), 7	2.59, ±0.18	(k) - (d)	(w), (b), (c), (d), (e), (k)³
36	Aniline	62-53-3	2.55	(1.41....2.41), 7	1.65, ±0.34	(z) - (f)	(z), (w), (c), (e), (f), (h), (q)
37	Anthracene	120-12-7	3.65	(4.27....4.42), 7	4.35, ±0.06	(c) - (h)	(w), (b), (c), (e), (f), (h), (v)
38	Atrazine	1912-24-9	2.80	(1.84....3.28), 11	2.35, ±0.35	(k) - (k)	(w), (c), (d), (e), (f), (h)², (k)³, (r)
39	Benzaldahyde	100-52-7	2.74	n.a.	n.a.	n.a.-n.a.	n.a.
40	Benzene	71-43-2	1.73	(1.11....2.45), 11	1.80, ±0.33	(k) - (k)	(w), (b), (c), (e), (f), (h), (k)³, (q), (v)

Chapter 9 - Appendix

41	Benzo(a)pyrene	50-32-8	5.43	(5.61...6.85), 5	6.09, ±0.48	(k) - (k)	(w), (b), (k)³
42	Fluoranthene	206-44-0	4.44	(4.48...5.24), 7	4.77, ±0.24	(k) - (k)	(w), (c), (e), (k)³, (r)
43	Fluorene	86-73-7	3.28	(3.68...3.70), 6	3.69, ±0.01	(f) - (w)	(w), (b), (c), (e), (f), (h)
44	Fluorobenzene	462-06-6	2.31	1	4.75	n.a.- n.a.	(v)
45	Hexachlorobenzene	118-74-1	4.45	(3.59...5.90), 7	4.50, ±1.00	(b) - (k)	(w), (b), (c), (d), (e), (h), (k)³
46	Lindane	58-89-9	3.79	(2.60...3.56), 9	3.03, ±0.24	(k) - (k)	(w), (b), (c), (d), (e), (h), (k)³
47	Naphthalene	91-20-3	3.23	(1.98...3.44), 10	2.90, ±0.38	(k) - (k)	(z), (w), (c), (e), (f), (h), (k)³, (v)
48	PCB 180	35065-29-3	6.52	n.a.	n.a.	n.a.- n.a.	n.a.
49	PCB 202	2136-99-4	6.16	(6.36...7.34), 3	6.69, ±0.57	(w) - (b)	(w), (b), (e)
50	Pentachlorobenzene	608-93-5	4.32	(2.93...5.53), 9	4.20, ±0.90	(k) - (k)	(w), (b), (c), (f), (h)², (k)³
51	Pentachlorophenol	87-86-5	2.74	(2.83...4.59), 12	3.74, ±0.78	(k) - (f)	(z), (w), (b), (c), (e), (f), (h)², (k)³, (v)
52	Phenol	108-95-2	2.49	(0.99...1.90), 9	1.46, ±0.25	(k) - (w)	(z), (w), (c), (e), (h), (k)³, (q)
53	Prometryne	7287-19-6	2.89	(1.99...3.15), 7	2.67, ±0.39	(k) - (d)	(w), (c), (d), (k)³, (r)
54	Propazine	139-40-2	3.03	(1.91...2.56), 9	2.36, ±0.21	(k) - (f)	(w), (c), (d), (e), (f), (h), (k)³
55	Pyridine	110-86-1	1.35	1	1.60	n.a.- n.a.	(w)
56	Sebuthylazine	7286-69-3	2.89	n.a.	n.a.	n.a.- n.a.	n.a.
57	Simetryn	1014-70-6	2.94	1	2.54	n.a.- n.a.	(w)
58	Terbuthylazine	5915-41-3	3.10	1	2.32	n.a.- n.a.	(c)
59	Terbutryn	886-50-0	2.86	(2.85...4.30), 7	3.32, ±0.55	(w) - (k)	(w), (c), (d), (e), (k)³
60	Toluene	108-88-3	2.43	(1.29...2.39), 11	2.00, ±0.29	(k) - (c)	(w), (b), (c), (e), (f), (h), (k)³, (q), (v)
61	Triclosan	3380-34-5	4.71	n.a.	n.a.	n.a.- n.a.	n.a.

n.a. is not available. The symbols of references are shown between brackets, and power numbers are the numbers of data that taken from the reference.

No.	Substances	CAS No.	Exp. log K_{oc} Aldrich humic acid in the current study	Exp. log K_{oc} (Soils & Sediments) in literature			References
				Range, n	Mean value of Exp. log K_{oc} ±SD	Min - Max	All
1	1,2,3,4-Tetrachlorobenzene	634-66-2	4.50	(3.28,...4.90), 9	3.82, ±0.49	(x) - (k)	(g)[2], (j), (k)[3], (u), (x)[2]
2	1,2,4-Trimethylbenzene	95-63-6	3.38	1	1.90	n.a.- n.a.	(k)
3	1,3,5-Trichlorobenzene	108-70-3	3.02	(3.26,...4.20), 3	3.81, ±0.49	(k) - (k)	(k)[3]
4	1-Aminopyrene	1606-67-3	4.82	n.a.	n.a.	n.a.- n.a.	n.a.
5	1-Nitronaphthalene	86-57-7	3.00	1	4.70	n.a.- n.a.	(m)
6	2,3,4,6-Tetrachlorophenol	58-90-2	2.90	(2.19,...3.89), 6	3.14	(k) - (a)	(a)[3], (k)[3]
7	2,3,5,6-Tetrachloraniline	3481-20-7	3.81	n.a.	n.a.	n.a.- n.a.	n.a.
8	2,3-Dimethylnaphthalene	581-40-8	3.68	n.a.	n.a.	n.a.- n.a.	n.a.
9	2,4,6-Trichlorophenol	88-06-2	2.37	(1.96,...3.12), 10	2.90, ±0.35	(k) - (a)	(a)[3], (g)[4], (k)[3]
10	2,4'-DDD	53-19-0	5.57	n.a.	n.a.	n.a.- n.a.	n.a.
11	2,4'-DDE	3424-82-6	5.41	n.a.	n.a.	n.a.- n.a.	n.a.
12	2,4'-DDT	789-02-6	5.13	n.a.	n.a.	n.a.- n.a.	n.a.
13	2,4-Dichlorophenol	120-83-2	2.60	(2.23,...2.85), 15	2.61, ±0.20	(k) - (a)	(a)[3], (g)[4], (k)[3], (p)[5]
14	2,4-Di-t-butylphenol	96-76-4	3.99	n.a.	n.a.	n.a.- n.a.	n.a.
15	2-Aminoanthracene	613-13-8	4.75	n.a.	n.a.	n.a.- n.a.	n.a.
16	2-Fluorobiphenyl	321-60-8	3.81	n.a.	n.a.	n.a.- n.a.	n.a.
17	2-Methylanthracene	613-12-7	4.50	n.a.	n.a.	n.a.- n.a.	n.a.
18	2-Methylpyridine	109-06-8	2.88	n.a.	n.a.	n.a.- n.a.	n.a.
19	2-Nitrofluorene	607-57-8	3.42	n.a.	n.a.	n.a.- n.a.	n.a.
20	3-Phenylphenol	580-51-8	3.33	n.a.	n.a.	n.a.- n.a.	n.a.

Chapter 9 - Appendix

21	4,4'-DDD	72-54-8	5.02	n.a.	n.a.- n.a.	n.a.
22	4,4'-DDE	72-55-9	5.02	1	5.05	(j)
23	4,4'-DDT	50-29-3	5.45	(3.85,...5.95), 6	5.01, ±0.86	(j) - (k)	(j)³, (k)³
24	4,4'-Isopropylidenediphenol	80-05-7	3.44	n.a.	n.a.	n.a. - n.a.	n.a.
25	4-Chloroaniline	106-47-8	3.61	(2.14,...2.80), 5	2.42, ±0.24	(p) - (p)	(p)⁵
26	4-Chlorophenol	106-48-9	2.96	(1.44...3), 14	2.07, ±0.40	(x) - (m)	(a), (k)³, (m), (p)⁵ (u), (x)³
27	4-n-Hexylphenol	2446-69-7	3.98	n.a.	n.a.	n.a.- n.a.	n.a.
28	4-Nitrophenol	100-02-7	2.62	(2.04,...2.72), 3	2.29, ±0.37	(k) - (k)	(k)³
29	4-n-Nonylphenol	104-40-5	4.87	n.a.	n.a.	n.a.- n.a.	n.a.
30	4-n-Octylphenol	1806-26-4	3.98	n.a.	n.a.	n.a.- n.a.	n.a.
31	6-Aminochrysene	2642-98-0	5.34	n.a.	n.a.	n.a.- n.a.	n.a.
32	6-Nitroquinoline	613-50-3	3.25	n.a.	n.a.	n.a.- n.a.	n.a.
33	Acridine	260-94-6	3.63	(3.74,...4.34), 5	4.14, ±0.24	(g) - (g)	(g)⁴, (l)
34	Aldrin	309-00-2	5.23	n.a.	n.a.	n.a.- n.a.	n.a.
35	Ametryne	834-12-8	2.70	(2.23,...2.59), 3	2.38, ±0.19	(k) - (k)	(k)³
36	Aniline	62-53-3	2.55	(1.28...2.38), 5	1.75, ±0.48	(p) - (p)	(p)⁵
37	Anthracene	120-12-7	3.65	(3.41...6.29), 16	4.65, ±0.72	(k) - (o)	(g)³, (j)³, (j)³, (k)³, (l), (o), (y)²
38	Atrazine	1912-24-9	2.80	(1.61...3.10), 10	2.19, ±0.44	(p) - (j)	(j), (k)³, (p)⁵, (u)
39	Benzaldahyde	100-52-7	2.74	n.a.	n.a.	n.a.- n.a.	n.a.
40	Benzene	71-43-2	1.73	(1.12,...2.34), 13	1.62, ±0.34	(x) - (m)	(g)⁴, (j), (k)³, (m), (u), (x)³
41	Benzo(a)pyrene	50-32-8	5.43	(4.00,...7.71), 12	5.76, ±1.26	(j) - (y)	(j)³, (j)³, (k)³, (o), (y)²
42	Fluoranthene	206-44-0	4.44	(3.80,...6.50), 15	5.16, ±0.87	(j) - (m)	(g)⁴, (j)³, (k)³, (m), (o), (y)³
43	Fluorene	86-73-7	3.28	(3.30,...5.49), 6	4.62, ±0.96	(j) - (j)	(j), (j), (k), (l), (o), (u)
44	Fluorobenzene	462-06-6	2.31	(4.14,...4.28), 2	4.21, ±0.10	(j) - (j)	(j)²
45	Hexachlorobenzene	118-74-1	4.45	(3.52,...5.50), 5	4.54, ±0.93	(j) - (k)	(j)², (k)³
46	Lindane	58-89-9	3.79	(1.50...4.16), 7	2.66, ±0.98	(j) - (m)	(j)², (j), (k)³, (m)
47	Naphthalene	91-20-3	3.23	(2.22,...4.87), 21	3.13, ±0.81	(x) - (j)	(g)⁴, (j)³, (k)³, (l), (m), (o), (u), (x)⁵

Chapter 9 - Appendix

#	Name	CAS				
48	PCB 180	35065-29-3	6.52	(5.50....6.40), 4	5.95, ±0.39	(k) - (j)	(j) -(k)³
49	PCB 202	2136-99-4	6.16	(5.21....6.36), 7	5.94, ±0.51	(j) - (g)	(g)², (j)², (k)³
50	Pentachlorobenzene	608-93-5	4.32	1	4.70	n.a. - n.a.	(j)
51	Pentachlorophenol	87-86-5	2.74	(2.60....4.59), 13	3.84, ±0.74	(a) - (a)	(a)³, (g)⁴, (j)³, (k)³
52	Phenol	108-95-2	2.49	(0.67....2.08), 20	1.38, ±0.37	(x) - (m)	(a)³, (k)³, (m), (p)¹⁵, (s)³, (x)⁵
53	Prometryne	7287-19-6	2.89	(1.82....4.06), 10	2.86, ±0.69	(p) - (j)	(j)², (k)³, (p)⁵
54	Propazine	139-40-2	3.03	(1.93....2.58), 8	2.16, ±0.20	(k) - (p)	(k)³, (p)⁵
55	Pyridine	110-86-1	1.35	n.a.	n.a.	n.a. - n.a.	n.a.
56	Sebuthylazine	7286-69-3	2.89	n.a.	n.a.	n.a. - n.a.	n.a.
57	Simetryn	1014-70-6	2.94	n.a.	n.a.	n.a. - n.a.	n.a.
58	Terbuthylazine	5915-41-3	3.10	n.a.	n.a.	n.a. - n.a.	n.a.
59	Terbutryn	886-50-0	2.86	(2.50....3.55), 8	2.96, ±0.39	(p) - (p)	(k)³, (p)⁵, (u)
60	Toluene	108-88-3	2.43	(1.13....2.86), 16	1.86, ±0.41	(j) - (m)	(g)⁴, (j)², (k)³, (m), (u), (x)⁵
61	Triclosan	3380-34-5	4.71	1	4.02	n.a. - n.a.	(u)

n.a. is not available. The symbols of references are shown between brackets, and power numbers are the numbers of data that taken from the reference.

Exp. log K_{oc} (Humic acids) in Literature

No.	Substances	CAS No.	Exp. log K_{oc} Aldrich humic acid in the current study	Rang, n	Mean value of Exp. log K_{oc} ±SD	Min - Max	References All
1	1,2,3,4-Tetrachlorobenzene	634-66-2	4.50	1	4.15	n.a.- n.a.	(v)
2	1,2,4-Trimethylbenzene	95-63-6	3.38	1	2.54	n.a.- n.a.	(v)
3	1,3,5-Trichlorobenzene	108-70-3	3.02	n.a.	n.a.	n.a.- n.a.	n.a.
4	1-Aminopyrene	1606-67-3	4.82	n.a.	n.a.	n.a.- n.a.	n.a.
5	1-Nitronaphthalene	86-57-7	3.00	1	4.76	n.a.- n.a.	(m)
6	2,3,4,6-Tetrachlorophenol	58-90-2	2.90	(2.07 ... 2.91), 4	2.51, ±0.46	(n) - (t)	(n)³, (t)
7	2,3,5,6-Tetrachloraniline	3481-20-7	3.81	n.a.	n.a.	n.a.- n.a.	n.a.
8	2,3-Dimethylnaphthalene	581-40-8	3.68	n.a.	n.a.	n.a.- n.a.	n.a.
9	2,4,6-Trichlorophenol	88-06-2	2.37	(1.72 ... 2.79), 4	2.36, ±0.49	(n) - (t)	(n)³, (t)
10	2,4'-DDD	53-19-0	5.57	n.a.	n.a.	n.a.- n.a.	n.a.
11	2,4'-DDE	3424-82-6	5.41	n.a.	n.a.	n.a.- n.a.	n.a.
12	2,4'-DDT	789-02-6	5.13	n.a.	n.a.	n.a.- n.a.	n.a.
13	2,4-Dichlorophenol	120-83-2	2.60	1	2.50	n.a.- n.a.	(t)
14	2,4-Di-t-butylphenol	96-76-4	3.99	n.a.	n.a.	n.a.- n.a.	n.a.
15	2-Aminoanthracene	613-13-8	4.75	n.a.	n.a.	n.a.- n.a.	n.a.
16	2-Fluorobiphenyl	321-60-8	3.81	n.a.	n.a.	n.a.- n.a.	n.a.
17	2-Methylanthracene	613-12-7	4.50	n.a.	n.a.	n.a.- n.a.	n.a.
18	2-Methylpyridine	109-06-8	2.88	1	2.29	n.a.- n.a.	(v)

#	Compound	CAS					
19	2-Nitrofluorene	607-57-8	3.42	n.a.	n.a.	n.a. - n.a.	n.a.
20	3-Phenylphenol	580-51-8	3.33	n.a.	n.a.	n.a. - n.a.	n.a.
21	4,4'-DDD	72-54-8	5.02	n.a.	n.a.	n.a. - n.a.	n.a.
22	4,4'-DDE	72-55-9	5.02	1	5.45	(j)	n.a.
23	4,4'-DDT	50-29-3	5.45	(4.55 ... 5.88), 7	5.41, ±0.42	(k)-(j)	(j), (j)³, (k)³
24	4,4'-Isopropylidenediphenol	80-05-7	3.44	n.a.	n.a.	n.a. - n.a.	n.a.
25	4-Chloroaniline	106-47-8	3.61	n.a.	n.a.	n.a. - n.a.	n.a.
26	4-Chlorophenol	106-48-9	2.96	(1.71 ... 3.07), 3	2.36, ±0.68	(v)-(m)	(m), (t), (v)
27	4-n-Hexylphenol	2446-69-7	3.98	n.a.	n.a.	n.a. - n.a.	n.a.
28	4-Nitrophenol	100-02-7	2.62	(1.38 ... 1.92), 3	1.66, ±0.27	(n)-(n)	(n)³
29	4-n-Nonylphenol	104-40-5	4.87	n.a.	n.a.	n.a. - n.a.	n.a.
30	4-n-Octylphenol	1806-26-4	3.98	n.a.	n.a.	n.a. - n.a.	n.a.
31	6-Aminochrysene	2642-98-0	5.34	n.a.	n.a.	n.a. - n.a.	n.a.
32	6-Nitroquinoline	613-50-3	3.25	n.a.	n.a.	n.a. - n.a.	n.a.
33	Acridine	260-94-6	3.63	(4 ... 4.24), 2	4.12, ±0.17	(j)-(l)	(j), (l)
34	Aldrin	309-00-2	5.23	1	5.05	n.a. - n.a.	(j)
35	Ametryne	834-12-8	2.70	1	2.53	n.a. - n.a.	(j)
36	Aniline	62-53-3	2.55	1	1.54	n.a. - n.a.	(v)
37	Anthracene	120-12-7	3.65	(3.92 ... 5.52), 12	4.47, ±0.39	(k)-(j)	(j)³, (j)³, (k)³, (l), (t'), (v)
38	Atrazine	1912-24-9	2.80	(1.91 ... 2.91), 3	2.48, ±0.53	(k)-(k)	(k)³
39	Benzaldahyde	100-52-7	2.74	1	1.71	n.a.-n.a.	(v)
40	Benzene	71-43-2	1.73	(1.51 ... 1.86), 5	1.69, ±0.15	(k)-(m)	(j), (k)³, (m)
41	Benzo(a)pyrene	50-32-8	5.43	(5.18 ... 6.56), 9	6.06, ±0.48	(j)-(k)	(j)³, (j)³, (k)³
42	Fluoranthene	206-44-0	4.44	(6.09 ... 4.59), 7	5.02, ±0.53	(j)-(m)	(j)³, (k)², (m), (t')
43	Fluorene	86-73-7	3.28	(3.00 ... 4.86), 8	3.93, ±0.51	(j)-(j)	(j), (j)², (k)³, (l), (t')
44	Fluorobenzene	462-06-6	2.31	(4.57 ... 4.95), 2	4.76, ±0.27	(j)-(j)	(j)²
45	Hexachlorobenzene	118-74-1	4.45	(4.86 ... 5.89), 3	5.22, ±0.58	(v)-(k)	(k)², (v)

- 214 -

Chapter 9 - Appendix

#	Name	CAS	value	range	mean±sd	refs1	refs2
46	Lindane	58-89-9	3.79	(1.50...3.58), 8	2.68, ±0.64	(i)- (m)	(i)², (l), (k)³, (m), (v)
47	Naphthalene	91-20-3	3.23	(2.89...4.00), 12	3.26, ±0.37	(k)- (m)	(i)³, (l)², (k)³, (l), (m), (t), (v)
48	PCB 180	35065-29-3	6.52	(5.09...5.73), 3	5.52, ±0.37	(l)- (l)	(i)², (k)
49	PCB 202	2136-99-4	6.16	(4.99...6.53), 3	5.71, ±0.77	(l)- (l)	(l)³
50	Pentachlorobenzene	608-93-5	4.32	(2.98...4.33), 3	3.46, ±0.75	(i)- (v)	(i)², (v)
51	Pentachlorophenol	87-86-5	2.74	(2.78...4.63), 10	3.54, ±0.63	(n)- (l)	(i)³, (k)³, (l)³, (t)
52	Phenol	108-95-2	2.49	(1.42...2.43), 3	1.90, ±0.51	(k)- (v)	(k), (m), (v)
53	Prometryne	7287-19-6	2.89	1	2.30	n.a.- n.a.	(k)
54	Propazine	139-40-2	3.03	n.a.	n.a.	n.a.- n.a.	n.a.
55	Pyridine	110-86-1	1.35	n.a.	n.a.	n.a.- n.a.	n.a.
56	Sebuthylazine	7286-69-3	2.89	n.a.	n.a.	n.a.- n.a.	n.a.
57	Simetryn	1014-70-6	2.94	n.a.	n.a.	n.a.- n.a.	n.a.
58	Terbuthylazine	5915-41-3	3.10	n.a.	n.a.	n.a.- n.a.	n.a.
59	Terbutryn	886-50-0	2.86	n.a.	n.a.	n.a.- n.a.	n.a.
60	Toluene	108-88-3	2.43	(1.94...2.97), 8	2.28, ±0.33	(i)- (i)	(i)³, (l), (k)³, (m)
61	Triclosan	3380-34-5	4.71	n.a.	n.a.	n.a.- n.a.	n.a.

n.a. is not available. The symbols of references are shown between brackets, and power numbers are the numbers of data that taken from the reference.

References

(a) Shiu WY, Ma KC, Varhaníčková D, Mackay D. *1994.* Chlorophenols and alkylphenols: A review and correlation of environmentally relevant properties and fate in an evaluative environment. *Chemosphere* 29(6):1155-1224.

(b) Tao S, Piao H, Dawson R, Lu X, Hu H. *1999.* Estimation of organic carbon normalized sorption coefficient (K_{oc}) for soils using the fragment constant method. *Environ. Sci. Technol.* 33:2719-2725.

(c) Sabljić A, Güsten H, Verhaa H, Hermens J. *1995.* QSAR modelling of soil sorption. Improvements and systematic of log K_{oc} vs. log K_{ow} correlations. *Chemosphere* 31:4489-4514.

(d) Lohninger H.*1994.* Estimation of soil partition coefficients of pesticides from their chemical structure. *Chemosphere* 29(8):1611-1994.

(e) Schüürmann G, Ebert R-U, Kühne R. *2006.* Prediction of the sorption of organic compounds into soil from molecular structure. *Environ. Sci. Technol.* 40:7005-7011.

(f) Baker JR, Mihelcic JR, Luehrs DC, Hickey JP. *1997.* Evaluation of estimation methods for organic carbon normalized sorption coefficients. *Water Environment Research* 69(2): 136-145.

(g) Nguyen TH, Goss KU, Ball WL. *2005.* Poly parameter linear free energy relationships for estimating the equilibrium partition of organic compounds between water and the neutral organic matter in soils and sediments. Critical Review. *Environ. Sci. Technol.* 39(4):913-924.

(h) Doucette WJ. *2000.* Soil and sediments sorption coefficients. In: Boethling R S, Mackay D (eds), Handbook of property estimation methods for chemicals: environmental and health sciences. *CRC Press LLC*, USA. pp: 141-190.

(i) Niederer C, Schwarzenbach R P, Goss KU. *2007.* Elucidating differences in the sorption properties of 10 humic and fulvic acids for polar and nonpolar organic chemicals. *Environ. Sci. Technol.* 41: 6711-6717.

(j) Krop HB, van Noort PCM, Govers HAJ. *2001.* Determination and theoretical Aspects of the equilibrium between dissolved organic matter and hydrophobic organic micropollutants in water (K_{doc}). In: George WW (esd), *Rev Environ Contam and Toxicol.* 169:1-122.

(k) Delle Site A. *2001.* Factors Affecting Sorption of Organic Compounds in Neutral Sorbent/Water Systems and Sorption Coefficients for Selected Pollutants, A Review, *J. Phys. Chem. Ref. Data* 30 (1): 187-439.

(l) Jonassen KEN. *2003.* Determination of physico-chemical constants in sorption of polycyclic aromatic compounds to soil organic matter - investigated by use of soil organic matter HPLC columns. Ph D Thesis. *Roskilde University*, Denmark.

(m) Endo S. *2008.* Characterization of sorption mechanisms to soil organic phases using molecular probe and polyparamter linear free energy relationship Approaches. PhD Thesis. *Eberhard Karls University*, Tübingen.

(n) Tülp H, Fenner K, Schwarzenbach RP, Goss KU.*2009.* pH-dependent Sorption of acidic organic chemicals to soil organic matter. *Environ. Sci. Technol.* 43: 9189-9195.

(o) Hawthorne SB, Grabanski CB, Miller DJ. *2006.* Measured partitioning coefficients for parent and alkyl polycyclic aromatic hydrocarbons in 114 historically contaminated sediments: Part1: K_{oc} values. *Environmental Toxicology and Chemistry* 25(11):2901-2911.

(p) Gawlik BM, Kettrup A, Muntau H. **2000**. Estimation of soil adsorption coefficients of organic compounds by HPLC screening using the second generation of the European references soil set. *Chemosphere* 41: 1337-1347.

(q) Hong H, Wang L, Han S. **1996**. Prediction adsorption coefficients (K_{oc}) for aromatic compounds by HPLC retention factors (k'). *Chemosphere* 32: 343-351.

(r) Szab G, Guczi J, Kördel W, Zsolnay A, Major V, Keresztes P. **1999**. Comparison of different HPLC stationary phases for determination of Soil-water distribution coefficient, K_{oc} values of organic chemicals in RP-HPLC system. *Chemosphere* 39(3): 431-442.

(s) Fiore S, Zanetti MC. **2009**. Sorption of phenols: influence of groundwater pH and of soil organic carbon content. *Am J. of Environ.Sci.* 5(4): 546-554.

(t) Ohenbusch G, Kumke UM, Frimmel HF. **2000**. Sorption of phenols to dissolved organic matter investigated by solid phase micro extraction. *Sci Total Environ* 253: 63-74.

(t') Kopinke FD, Pörschmann J, Georgi A. **1999**. Application of SPME to study sorption phenomena on dissolved humic organic matter. In: Pawliszyn J (eds), Application of solid phase microextraction. *The Royal Society of Chemistry, Thomas Graham House, Cambridge, UK*, pp: 111-127.

(u) Bronner G, Goss KU. **2011a**. Prediction sorption of pesticides and other multifunctional organic chemicals to soil organic carbon. *Environ.Sci. Technol.* 45: 1313-1319.

(v) Niederer C, Goss KU, Schwarzenbach RP. **2006**. Sorption Equilibrium of a Wide Spectrum of Organic Vapors in Leonardite Humic Acid: Modeling of Experimental Data. *Environ. Sci. Technol.* 40 (17): 5374–5379.

(w) Epi suite™v.4. U.S Environmental Protection Agency **2008**. Meylan WM, EPSUITE, Syracuse Research Corporation, Syracuse, NY. *http://www.syrres.com/esc/est_soft.htm*.

(x) Bronner G, Goss KU. **2011b**. Sorption of Organic Chemicals to Soil Organic Matter: Influence of Soil Variability and pH Dependence. *Environ. Sci. Technol.* 45: 1307-1312

(y) Ter Laak LT. **2005**. Sorption to soil of hydrophobic and ionic organic compounds: measurement and modeling. PhD Thesis. *Utrecht University*, Netherlands.

(z) Schüürmann G, Kühne R, Kleint F, Ebert RU, Rothenbacher C, Herth P. **1997**. „In-House software" ChemProp, Software system for automatic chemical property estimation from molecular structure. In: Chen F, Schüürmann G (eds), In Quantitative Structure-Activity Relationships in Environmental Sciences-VII. *SETAC Press: Pensacola*: 93-114.

- 217 -

9.17 The Calculated Values of $\log K_{OC}$ from Different Prediction Methods

Substances	CAS No.	Calc. log K_{OC} (2D molecular structure model)[a]	Calc. log K_{OC} (Epi suite™ v.4 from log K_{OW})[b]	Calc. log K_{OC} (Epi suite™ v.4 molecular topology)[c]	Calc. log K_{OC} (Molecular connectivity indices)[d]	Calc. log K_{OC} (Fragment constant method)[e]	LSER Poole (Exp. descriptors)[f]	LSER Poole (Calc. descriptors Platts method)[f]	LSER Nguyen (Exp. descriptors)[g]	LSER Nguyen (Calc. descriptors Platts method)[g]
1,2,3,4-Tetrachlorobenzene	634-66-2	3.64	3.61	3.35	3.10	3.50	3.60	3.51	3.47	3.37
1,2,4-Trimethylbenzene	95-63-6	2.60	3.15	2.79	2.88	3.06	2.66	2.87	2.70	2.87
1,3,5-Trichlorobenzene	108-70-3	3.30	3.39	3.12	2.87	3.11	3.20	3.26	3.09	3.08
1-Aminopyrene	1606-67-3	4.87	3.49	4.94	3.39	4.74	n.a.	4.21	n.a.	4.38
1-Nitronaphthalene	86-57-7	3.03	2.91	3.39	2.68	3.06	3.37	3.21	2.96	2.86
2,3,4,6-Tetrachlorophenol	58-90-2	3.74	3.70	3.47	3.70	3.14	3.17	3.34	3.43	3.69
2,3,5,6-Tetrachloroaniline	3481-20-7	3.53	3.91	3.87	3.39	3.55	3.70	3.29	3.59	3.44
2,3-Dimethylnaphthalene	581-40-8	3.41	3.82	3.61	3.70	3.81	3.47	3.62	3.65	3.66
2,4,6-Trichlorophenol	88-06-2	3.38	3.28	3.25	3.22	2.88	2.91	2.92	3.13	3.15
2,4-Dichlorophenol	120-83-2	3.00	2.78	2.70	2.83	2.50	2.55	2.39	2.60	2.59
2,4-Di-t-butylphenol	96-76-4	3.43	3.96	3.95	3.53	3.53	3.53	3.46	4.06	4.22
2-Aminoanthracene	613-13-8	4.20	2.98	4.42	3.17	4.24	4.51	3.67	4.38	3.83
2-Fluorobiphenyl	321-60-8	3.01	3.44	3.92	4.02	3.71	n.a.	3.49	n.a.	3.40
2-Methylanthracene	613-12-7	4.22	4.34	4.42	4.51	4.55	4.53	4.38	n.a.	4.45
2-Methylpyridine	109-06-8	1.41	1.72	2.06	1.60	1.71	1.27	1.45	0.91	1.23
2-Nitrofluorene	607-57-8	3.33	3.00	4.15	2.77	3.45	4.78	4.04	3.76	3.78
3-Phenylphenol	580-51-8	3.24	2.88	3.82	2.70	3.45	3.05	2.92	3.25	3.19
4,4'-Isopropylidenediphenol	80-05-7	3.36	3.10	4.58	2.75	3.18	2.92	3.08	3.67	3.91
4-Chloroaniline	106-47-8	2.44	1.92	2.05	1.98	2.40	2.07	1.95	1.99	1.96
4-Chlorophenol	106-48-9	2.64	2.41	2.48	2.40	2.11	2.10	1.67	2.08	1.93
4-Hexylphenol	2446-69-7	3.17	3.59	3.80	3.37	3.12	n.a.	3.09	n.a.	3.57

Chapter 9 - Appendix

Compound	CAS								
4-Nitrophenol	100-02-7	2.17	2.37	2.46	2.10	1.84	2.14	1.53	1.77	1.55
4-n-Nonylphenol	104-40-5	3.62	4.28	4.58	3.81	3.82	n.a.	3.97	n.a.	4.65
4-Octylphenol	1806-26-4	3.46	4.13	4.32	3.88	3.71	3.75	3.68	4.31	4.29
6-Aminochrysene	2642-98-0	5.14	4.33	5.47	3.61	5.22	n.a.	4.76	n.a.	5.01
6-Nitroquinoline	613-50-3	3.02	2.15	3.38	1.97	2.44	n.a.	2.45	n.a.	2.02
Acridine	260-94-6	3.67	2.95	4.21	2.80	3.44	3.59	3.29	3.86	3.21
Aldrin	309-00-2	4.52	5.26	4.91	5.00	4.74	n.a.	4.73	n.a.	4.28
Ametryne	834-12-8	2.64	2.36	2.63	2.49	2.36	2.69	2.09	3.18	2.11
Aniline	62-53-3	2.07	1.40	1.85	1.49	2.01	1.41	1.51	1.52	1.46
Anthracene	120-12-7	4.08	3.86	4.21	4.30	4.05	4.31	4.05	4.43	4.05
Atrazine	1912-24-9	2.49	2.16	2.35	2.32	2.29	2.45	1.96	2.44	1.77
Benzaldehyde	100-52-7	1.98	1.51	1.04	1.79	n.a.	1.76	1.54	1.44	1.22
Benzene	71-43-2	2.18	1.85	2.16	2.26	1.83	1.84	1.88	1.67	1.68
Benzo(a)Pyrene	50-32-8	5.70	5.32	5.77	5.86	5.67	6.14	5.67	6.47	5.80
Fluoranthene	206-44-0	4.22	4.48	4.74	4.84	4.57	4.74	4.59	4.76	4.61
Fluorene	86-73-7	3.42	3.63	3.96	4.05	3.33	3.65	3.80	3.71	3.78
Fluorobenzene	462-06-6	2.05	1.97	2.37	2.46	1.99	1.87	1.86	1.61	1.58
Hexachlorobenzene	118-74-1	4.37	4.24	3.79	3.54	4.27	4.34	4.63	4.37	4.60
Lindane	58-89-9	3.27	3.59	3.45	3.54	3.15	n.a.	3.66	n.a.	3.80
Naphthalene	91-20-3	3.12	2.86	3.19	3.28	2.94	3.02	2.97	2.96	2.87
2,4'-DDD	53-19-0	4.54	5.09	5.08	5.17	4.77	n.a.	5.46	n.a.	5.65
2,4'-DDE	3424-82-6	4.53	5.21	5.08	5.17	4.99	n.a.	5.59	n.a.	5.60
2,4'-DDT	789-02-6	4.73	5.89	5.23	5.32	5.40	n.a.	5.74	n.a.	5.88
4,4'-DDD	72-54-8	4.54	5.22	5.07	5.16	4.77	n.a.	5.46	n.a.	5.65
4,4'-DDE	72-55-9	4.53	5.65	5.07	5.16	4.99	n.a.	5.59	n.a.	5.60
4,4'-DDT	50-29-3	4.73	6.00	5.23	5.31	5.40	5.27	5.74	5.59	5.88
PCB 180	35065-29-3	5.74	5.64	5.54	5.29	6.25	6.26	6.42	6.20	6.45

Compound	CAS									
PCB 202	2136-99-4	6.09	5.34	5.77	6.63	6.69	7.10	6.62	n.a	7.08
Pentachlorobenzene	608-93-5	4.01	3.93	3.57	3.32	3.88	3.95	4.10	3.96	3.99
Pentachlorophenol	87-86-5	4.10	4.07	3.69	3.68	3.65	3.62	3.84	4.02	4.33
Phenol	108-95-2	2.27	1.90	2.27	1.82	1.73	1.56	1.29	1.61	1.38
Prometryne	7287-19-6	2.80	2.66	2.82	2.74	2.51	2.98	2.33	3.54	2.45
Propazine	139-40-2	2.65	2.33	2.54	2.47	2.44	2.64	2.20	2.70	2.12
Pyridine	110-86-1	1.27	1.46	1.86	1.36	1.22	1.02	1.13	0.62	0.84
Sebuthylazine	7286-69-3	2.62	2.54	2.63	2.41	2.52	n.a.	2.26	n.a	2.13
Simetryn	1014-70-6	2.49	2.26	2.45	2.41	2.03	2.47	1.86	2.90	1.76
Terbuthylazine	5915-41-3	2.49	2.60	2.19	2.10	2.50	2.80	2.19	2.96	2.10
Terbutryn	886-50-0	2.78	2.85	2.32	2.43	2.57	3.04	2.32	3.56	2.43
Toluene	108-88-3	2.37	2.46	2.21	2.08	2.33	2.13	2.21	2.02	2.08
Triclosan	3380-34-5	3.93	3.50	4.21	4.47	4.23	4.47	4.21	4.76	4.47

a. Schüürmann G, Ebert R-U, Kühne R. *2006*. Prediction of the sorption of organic compounds into soil from molecular structure. *Environ. Sci. Technol.* 40:7005-7011.

b. Meylan WM. *2000*. PCKOCWIN 1.66. *Syracuse Research Corporation, Syracuse, NY*;Meylan WM. EPSUITE. *Syracuse Research Corporation, Syracuse, NY*.http://www.syrres.com/esc/est_soft.htm

c. Meylan WM, Howard PH, Boethling RS. *1992*. Molecular topology/fragment contribution method for prediction soil sorption coefficients. *Environ. Sci. Technol.* 26:1560-1567.

d. Sabljić A, Güsten H, Verhaa H, Hermens J. *1995*. QSAR modelling of soil sorption. Improvements and systematic of log K_{oc} vs. log K_{ow} correlations. *Chemosphere* 31:4489-4514.

e. Tao S, Piao H, Dawson R, Lu X, Hu H.*1999*. Estimation of organic carbon normalized sorption coefficient (K_{oc}) for soils using the fragment constant method. *Environ. Sci. Technol.* 33:2719-2725.

f. Poole SK, Poole CF. *1999*. Chromatographic models for the sorption of neutral organic compounds by soil from water and air. *J. of Chromatography A* 845: 381-400.

g. Nguyen TH, GossKU, Ball WL. *2005*. Poly parameter linear free energy relationships for estimating the equilibrium partition of organic compounds between water and the neutral organic matter in soils and sediments. *Critical Review. Environ. Sci. Technol.* 39(4): 913-924.

9.18 The Used Equations of Franco and Trapp Model

The new prediction model of soil sorption of organic electrolytes based on pK_a was proposed by the Franco and Trapp. That theoretical model depends on the log $K_{ow(neutral\ species)}$ and pK_a of selected solutes. It can be used to predict the soil sorption of ionizable substances in the soil and sediments. In the chapter 4-2, the following equations were used for comparison only with weak electrolyte, weak acids, and weak base models. The octanol–water partition coefficient of the neutral molecule (P_n) but in the current study log K_{ow} was used.

For weak electrolyte models (neutral fractions) $\quad \log K_{oc} = 0.50 \log P_n + 1.13 \quad$ (9.1)

For weak acids log K_{oc} = 0.54 log P_n + 1.11 (9.2)

For weak bases log K_{oc} = 0.42 log P_n + 1.34 (9.3)

But the model has limitation for bases at pH 4.5 only and with wrong calculation of acids as explained in the following paragraph:

ChemProp Manual 1.8.5:

For acids, the model according to Franco et al. (2009) will be applied. This is the default behavior. The application of the base model from the 2008 paper can be forced in the parameters settings. In this case, due to model limitation the pH input will not be considered. A fix pH of 4.5 will be applied instead. Supplying of a valid pK_a is required: the models do not work for neutral compounds without specification of pK_a. Knowledge of the K_{ow} of the neutral species (P_n) is required. This is usually not available from experimental data, but prediction models typically estimate the K_{ow} of the neutral species P_n instead of the effective K_{ow} that is available from measurement. Alternatively, manual input of P_n (i.e., K_{ow} of the neutral species) can be selected.

For acids

$$K_{oc} = \frac{10^{0.54 \log P_n + 1.11}}{1 + 10^{pH - 0.6\ pK_a}} + \frac{10^{0.11 \log P_n + 1.54}}{1 + 10^{pK_a - pH}} + 0.6 \quad (9.4)$$

By substitution $pK_a \simeq 50$ in equation (9.4) for neutral solutes (unionized fractions), then equation (9.2) will be obtained.

For bases K_{oc} at pH=4.5 is calculated by

$$K_{oc} = (\varphi_n \ 10^{0.37\ (\log P_n + 1.70)} + \varphi_{ion}\ 10^{pK_a^{0.65} f^{0.14}}) \quad (9.5)$$

With Handerson-Hasselbalch for bases $\phi_n = 1 - \phi_{ion}$

$$\phi_{ion} = \left[\frac{1}{1 + 10^{pK_a - 4.5}} \right] \quad (9.6)$$

With,

$$f = \frac{K_{ow,7}}{K_{ow,7} + 1} \quad (9.7)$$

$$K_{ow,7} = \frac{P_n}{1 + 10^{pK_a - 7}} + \frac{P_{ion}}{1 + 10^{7 - pK_a}} \quad (9.8)$$

and $\log P_{ion} = P_n - 3.5 \quad (9.9)$

Kühne R, Ebert R-U, Schüürmann G. **2011**.Manual Implemented Models.*Department of Ecological Chemistry, Helmholtz Centre for Environmental Research- UFZ,Leipzig, Germany*: ChemProp manual 1.8.5.

Franco A, Trapp S.**2008**. Estimation of the soil-water partition coefficient normalized to organic carbon for ionisable organic chemicals. *Environmental Toxicology and Chemistry.*27 (10): 1995-2004.

Franco A, Fu W, Trapp S.**2009**. Influence of soil pH on the sorption of ionizable chemicals: Modeling Advances. *Environmental Toxicology and Chemistry* 28(3): 458-464.

9.19 The Derivation of K_{oc} Equation

SPME experiments are performed in closed system with definite volume V_m and definite gas volume V_{gas}. The mass balance of the distribution of the analyte mass between different phases in humic acid system can be expressed as the following equation:

$$m_m^{Dom} = m_{m,F}^{Dom} + m_{m,w}^{Dom} + m_{m,Dom}^{Dom} + m_{m,gas}^{Dom} \quad (9.10)$$

The total of analyte mass in humic acid system = The analyte mass in PDMS + The analyte mass in water phase + The analyte mass in humic acid + The analyte mass in gas

$$K_h = \frac{C_{gas}}{C_{water}}$$

$$m_{gas} = m_{water} \longrightarrow m_{gas} = C_g V_g = m_{water} = C_w V_w \longrightarrow V_w = \frac{C_g V_g}{C_w} = K_h V_g \quad (9.11)$$

$$m_{gas} = m_{water} \longrightarrow \frac{m_{gas}}{m_w} V_w = K_h V_g \longrightarrow m_{gas} = \frac{K_h V_g}{V_w} m_w \quad (9.12)$$

$$K_{oc} = \frac{C_{OC}}{C_{water}}$$

$$K_{oc} C_{water} = C_{OC} = K_{OC} \frac{V_w}{V_m} \frac{m_w}{m_w} \frac{V_m}{m_w} \longrightarrow \frac{m_w}{m_w} V_w = K_{OC} V_w$$

$$C_{OC} \frac{V_w}{m_w} = K_{OC} \frac{m_w}{m_w} \frac{V_m}{m_w} \longrightarrow \frac{m_{Dom}}{m_w} V_w = K_{OC} m_{OC} \quad (9.13)$$

$$C_{OC} \frac{V_w}{m_w} = K_{OC} \longrightarrow m_{Dom} = \frac{K_{OC} m_{OC}}{V_w} \cdot m_w$$

$$m_m^{Dom} = m_{m,F}^{Dom} + m_{m,w}^{Dom} + m_{m,Dom}^{Dom} + m_{m,gas}^{Dom} \longrightarrow m_m^{Dom} = m_{m,F}^{Dom} + m_{m,w}^{Dom} + \frac{K_{OC} m_{OC}}{V_w} \cdot m_w + \frac{K_h V_g}{V_w} m_w \quad (9.14)$$

- 223 -

Chapter 9 - Appendix

$$m_m^{Dom} = m_{m,F}^{Dom} + m_{m,w}^{Dom}\left(1 + \frac{K_{OC}m_{OC}}{V_w} + \frac{K_h V_g}{V_w}\right) \quad (9.15)$$

$$K_f = \frac{C_{fibre}}{C_{water}} = \frac{V_f}{\frac{m_w}{V_w}} = \frac{m_f}{V_w} \quad \longrightarrow \quad \frac{K_f m_w}{m_f} = \frac{m_f}{V_f m_f} \quad \longrightarrow \quad m_w = \frac{V_w}{V_f K_f} m_f \quad (9.16)$$

$$m_m^{Dom} = m_{m,F}^{Dom} + m\left(1 + \frac{K_{OC}m_{OC}}{V_w} + \frac{K_h V_g}{V_w}\right) \quad (9.17)$$

$$m_m^{Dom} = m_{m,F}^{Dom}\left(\frac{1+V_m}{V_f K_f}\right) + \frac{m_f K_{OC} m_{OC}}{V_f K_f} + \frac{m_f K_h V_g}{V_f K_f} \quad (9.18)$$

$$m_m^{Dom} = m_{m,F}^{Dom} \frac{V_f K_f + V_w + K_{OC} m_{OC} + K_h V_g}{K_f V_f} \quad (9.20)$$

With the same concept the mass balance of the distribution of the analyte mass between different phases in stander solution system can be expressed as the following equation:

$$m_m^{St} = m_{m,F}^{St} + m_{m,w}^{St} + m_{m,gas}^{St} \quad (9.21)$$

$$m_{gas} = \frac{K_h V_g}{V_w} m_w \quad \longrightarrow \quad m_m^{St} = m_{m,F}^{St} + m_{m,w}^{St} + \frac{K_h V_g}{V_w} m_w \quad (9.22)$$

$$m_m = \frac{V_m}{V_f K_f} m_f \quad \longrightarrow \quad m_m^{St} = m_{m,F}^{St} + \frac{V_m m_f}{K_h V_g} + \frac{K_h V_g}{V_w} \frac{V_m m_f}{K_f V_f} \quad \longrightarrow \quad m_m^{St} = m_{m,F}^{St}\left(1 + \frac{V_m}{K_f V_f} + \frac{K_h V_g}{K_f V_f}\right) \quad (9.23)$$

$$m_m^{St} = m_{m,F}^{St}\left(\frac{K_f V_f}{K_f V_f} + \frac{V_m}{K_f V_f} + \frac{K_h V_g}{K_f V_f}\right) \quad \longrightarrow \quad m_m^{St} = m_{m,F}^{St}\left(\frac{K_f V_f + V_w + K_h V_g}{K_f V_f}\right) \quad (9.24)$$

Chapter 9 - Appendix

By using the difference correction factor between the total analyte mass in humic acid system and standard solution system in SPME sampling

$$a = \left(\frac{m_{m,F}^{Dom}}{m_m^{St}}\right)_{gas} = \frac{m_{m,F}^{Dom}}{m_{m,F}^{St}} \left(\frac{V_f K_f + V_w + K_{oc} m_{humic} + K_h V_g}{K_f V_f}\right) \tag{9.25}$$

$$a \left(\frac{m_{m,F}^{St}}{m_{m,F}^{Dom}}\right) = \left(\frac{V_f K_f + V_w + K_{oc} m_{humic} + K_h V_g}{K_f V_f}\right)\left(\frac{K_f V_f}{K_f V_f + V_w + K_h V_g}\right) \tag{9.26}$$

$$a \left(\frac{m_{m,F}^{St}}{m_{m,F}^{Dom}}\right)[K_f V_f + V_w + K_h V_g] = [V_f K_f + V_w + K_{oc} m_{humic} + K_h V_g] \tag{9.27}$$

$$m_{oc} = m_{humic} = C_{humic} f_{oc} V_m \tag{9.28}$$

$$a \left(\frac{m_{m,F}^{St}}{m_{m,F}^{Dom}}\right)[K_f V_f + V_w + K_h V_g] - V_f K_f - V_w - K_h V_g = [K_{oc} m_{oc}] = K_{oc} C_{humic} f_{oc} V_m \tag{9.29}$$

$$a \left(\frac{m_{m,F}^{St}}{m_{m,F}^{Dom}} - 1\right)[K_f V_f + V_w + K_h V_g] = [K_{oc} m_{oc}] = K_{oc} C_{humic} f_{oc} V_m \tag{9.30}$$

$$a \left(\frac{m_{m,F}^{St}}{m_{m,F}^{Dom}} - 1\right)\left(\frac{1}{C_{humic} f_{oc}}\right)\left[\frac{K_f V_f + V_w + K_h V_g}{V_m}\right] = K_{oc} \tag{9.31}$$

It was proposed that the difference correction factor between analyte mass of both the humic acid and standard solution systems will be equal 1 at equilibrium and the same condition of sampling with the same used fibre volume

Chapter 9 - Appendix

$$a\left(\frac{m_{m,F}^{St}}{m_{m,F}^{Dom}}-1\right)\left(\frac{1}{C_{humic}f_{oc}}\right)\left[\frac{K_fV_f+K_hV_g}{V_m}+1\right]=K_{oc} \qquad (9.32)$$

$$\left(\frac{m_{m,F}^{St}-m_{m,F}^{Dom}}{m_{m,F}^{Dom}}\right)\left(\frac{1}{C_{humic}f_{oc}}\right)\left[\frac{K_fV_f+K_hV_g}{V_m}+1\right]=K_{oc}$$

$$\left(\frac{m_{m,F}^{St}V-m_{m,F}^{Dom}V_m}{m_{m,F}^{Dom}V_m}\right)\left(\frac{1}{C_{humic}f_{oc}}\right)\left[\frac{K_fV_f+K_hV_g}{V_m}+1\right]=K_{oc} \qquad (9.33)$$

$$\left(\frac{C^{St}V-C^{Dom}V_m}{C^{Dom}V_m}\right)\left(\frac{1}{C_{humic}f_{oc}}\right)\left[\frac{K_fV_f+K_hV_g}{V_m}+1\right]=K_{oc} \qquad (9.34)$$

$$\left(\frac{C^{St}_{ref}-C^{Test}_{matrix}}{C^{Test}_{matrix}}\right)\left(\frac{1}{C_{humic}f_{oc}}\right)\left[\frac{K_fV_f+K_hV_g}{V_m}+1\right]=K_{oc}$$

$$\left(\frac{C_{sorb(bound)}}{C_{free(diss)}}\right)\left(\frac{1}{C_{humic}f_{oc}}\right)\left[\frac{K_fV_f+K_hV_g}{V_m}+1\right]=K_{oc}$$

$$\left(\frac{A^{St}_{ref}-A^{Test}_{matrix}}{A^{Test}_{matrix}}\right)\left(\frac{1}{C_{humic}f_{oc}}\right)\left[\frac{K_fV_f+K_hV_g}{V_m}+1\right]=K_{oc}[L\,kg^{-1}] \qquad (9.35)$$

The above equations (9.10) and (9.21) from the following references:

Georgi A. **1998**. Sorption von hydrophoben organischen Verbindungen an gelösten Huminstoffen. *Dissertation in der UFZ- Leipzig.*

Kopinke FD, Pörschmann J, Georgi A. **1999**. Application of SPME to study sorption phenomena on dissolved humic organic matter. In: Pawliszyn J (eds), Application of solid phase microextraction. *The Royal Society of Chemistry: Thomas Graham House, Cambridge, UK.* pp: 111-127.

Mackenzie K, Georgi A, Kumke M, Kopinke FD. **2002**. Sorption of pyrene to dissolved humic substances and related model polymers. 2. Solid phase micro extraction (SPME) and fluorescence quenching technique (FQT) as analytical methods. *Environ. Sci. Technol.* 36: 4403-4409.

Poerschmann J, Kopinke FD and Pawliszyn J. **1997**. Solid phase microextraction for determining the distribution of chemicals in aqueous matrices. *Anal. Chem.* 69:597-600.

- 226 -

9.20 The Statistical Treatment of Experimental K_{oc} Data

1. Statistical Functions for Uncertainty in K_{oc} Measurement

The calculation of logarithmic Uncertainty value of the water-humic acid partitioning coefficient (log K_{oc}) has been done using Excel 2007 software. For example for anthracene at pH 7, we have 4 values of K_{oc} which calculated by calibration curve method I as mentioned in experimental part

Cell No.	K_{oc} (L/kg)
A1	4469
A2	5183
A3	4184
A4	4136

Note that in our selection of experimental values. The outliers values are the most deviated values from the mean value (highest maximum and lowest minimum values) should be excluded from our calculations. In the above example the number of measurements (sample size) n = 4, square root .of sample size = $\sqrt{4}$ = 2.

Mean value: The average (arithmetic mean) of the arguments.

$$\bar{x} = \frac{1}{n}\sum_{i=1}^{n} x_i \qquad (9.36)$$

Formula by Excel = AVERAGE (A1:A4) = 4493.550923

Log mean value = log $_{10}$(4493.550923) = 3.65

Stander deviation of the sample size: is a measure of how widely values are dispersed from the average value (the mean)

Formula by Excel = STDEV (A1:A4) = 482.9936964

$$s = \sqrt{\frac{1}{N-1}\sum_{i=1}^{N}(x_i - \bar{x})^2} \qquad (9.37)$$

The standard error of the mean (SEM): is the standard deviation of the sample mean estimate of a population mean. (It can also be viewed as the standard deviation of the error in the sample mean relative to the true mean, since the sample mean is an unbiased estimator.) SEM is usually estimated by standard deviation of the sample divided by the square root of the sample size (assuming statistical independence of the values in the sample):

$$SE_{\bar{x}} = \frac{s}{\sqrt{n}}$$ (9.38)

Where, s is the deviation and n is the sample size (number of items)

Standard error of mean = (482.9936964 / 2) = 241.4968482

Relative standard deviation: which is simply the standard deviation divided by the mean.

$$RSD = \frac{s}{\bar{x}} \quad RSD\,\% \text{ (Coefficient of variation) or the precision} = \frac{s}{\bar{x}}(100) = 10.74859737$$ (9.39)

95% Confidence Interval: The standard error can be used to calculate confidence intervals for the true population mean. For a 95% 2-sided confidence interval, the Upper Confidence Limit (UCL) and Lower Confidence Limit (LCL) are calculated as:

$$95\%ULC = Mean + 1.96 \ (StErr) = \bar{y} + 1.96 \frac{s}{\sqrt{n}}$$ (9.40)

$$95\%ULC = Mean - 1.96 \ (StErr) = \bar{y} - 1.96 \frac{s}{\sqrt{n}}$$ (9.41)

To get a 90% or 99% confidence interval, you would change the value 1.96 to 1.645 or 2.575, respectively. The value 1.96 represents the 97.5 percentile of the standard normal distribution. (You may often see this number rounded to 2). To calculate a different percentile of the standard normal distribution, you can use the NORMSINV function in Excel. **Example:** 1.96 = NORMSINV (1-(1-.95)/2).

Gaussian error in space measurements *(derived by Mr. Dominik Wondrousch)*

Relative maximum error = $R = \dfrac{Error_{Max}}{X_{mean}} = \overline{X} \pm Error_{Max}$ (9.42)

$$R = \left[\left| \dfrac{ULC - K_{oc(mean)}}{K_{oc(mean)}} \right| , \left| \dfrac{LLC - K_{oc(mean)}}{K_{oc(mean)}} \right| \right] \quad (9.43)$$

Formula in Excel is = ABS ((ULC- K_{oc}) / K_{oc}), ABS ((LLC- K_{oc}) / K_{oc})

Logarithmic function of error:

$\text{Log } K_{oc} \pm \left| -\log(1-R) \right|$ (9.44)

$\text{Log } (z+u) = u\,(\log) = \dfrac{\partial f}{\partial c} u = \dfrac{\partial \ln(c)}{\partial c \ln(10)} u = \dfrac{1}{\ln 10} \dfrac{1}{c} u$ 1st order (9.45)

$u(\log) = \dfrac{1}{\ln(10)} \sum_{n=1}^{\infty} \left| (-1)^{n-1} \dfrac{u}{nX^n} \right| = -\log\!\left(1 - \dfrac{u}{c}\right) = -\log(1-R)$ (9.46)

$\text{Log}\!\left(\dfrac{C_A + U_A}{C_B + U_B}\right) = \log(C_A + U_A) - \log(C_B + U_B)$ (9.47)

$u\,(\log K) = -\log(1-R_A) - \log(1-R_B)$ (9.48)

2. Error Propagation for log x *(derived by Prof. Gerrit Schüürmann and Mr. Dominik Wondrousch)*

A) Simple log x

Consider log x, where x may contain some error Δx:

$$\log(x + \Delta x) = \log\left[x\left(1 + \frac{\Delta x}{x}\right)\right] = \log x + \log\left(1 + \frac{\Delta x}{x}\right) \qquad (9.49)$$

Similarly,

$$\log(x - \Delta x) = \log\left[x\left(1 - \frac{\Delta x}{x}\right)\right] = \log x + \log\left(1 - \frac{\Delta x}{x}\right) \qquad (9.50)$$

Combination of equations (9.49) and (9.50) leads to

$$\log(x \pm \Delta x) = \log\left[x\left(1 \pm \frac{\Delta x}{x}\right)\right] = \log x + \log\left(1 \pm \frac{\Delta x}{x}\right) \qquad (9.51)$$

as formula to calculate the (unsymmetric) error in log x based on an error in x (assuming the latter to be symmetric).

B) Log x as ratio of two other variables

Consider now log x with $x = \dfrac{c_a}{c_b}$, such as log K_{oc} or log K_{mw}. In this more involved case, the error in x results from (possibly independent) errors in the nominator, c_a, and the denominator, c_b.

First, we analyze the function separately.

$$f = \frac{c_a}{c_b} = c_a \cdot \frac{1}{c_b} \qquad (9.52)$$

Its error Δf based on errors in c_a and c_b can be estimated through the Taylor expansion developed to (only) first order:

$$\Delta f = \frac{1}{c_b}\Delta c_a + c_a \frac{1}{c_b^2}\Delta c_b \qquad (9.53)$$

Note that in eq. (9.53), the 2nd term on the r.h.s. would have a negative sign through taking the derivative $[\frac{d}{dx}\frac{1}{x} = -\frac{1}{x^2}]$, which is however converted to a positive sign because errors add up and to not cancel. The relative error of f, $\frac{\Delta f}{f}$, results from dividing eq. (9.53) by f with f being defined through eq. (9.52):

$$\frac{\Delta f}{f} = \frac{\Delta c_a}{c_a} + \frac{\Delta c_b}{c_b}$$

Eq. (9.54) means that the individual relative errors add up to the total relative error. Returning now to log x with $x = \frac{c_a}{c_b}$, we can use eq. (9.54) with $f \equiv x$ to express the relative error in x, $\frac{\Delta x}{x}$, through the errors of its components c_a and c_b:

$$\log\left(\frac{c_a}{c_b} \pm \Delta\left[\frac{c_a}{c_b}\right]\right) = \log\frac{c_a}{c_b} + \log\left(1 \pm \left[\frac{\Delta c_a}{c_a} + \frac{\Delta c_b}{c_b}\right]\right) \qquad (9.55)$$

Eq. (9.55) thus yields the error in $\log\frac{c_a}{c_b}$ caused by errors in c_a and c_b, the latter of which are called Δc_a and Δc_b, respectively.

9.21 Statistical Parameters of Model Calibration with their Mathematical Formula.

Parameter	Definition	Mathematical formula		Statistics Program
r^2	squared correlation coefficients	$= 1 - \dfrac{\sum_{i=1}^{N}(y_{fit} - y_{exp})^2}{\sum_{i=1}^{N}(y_{exp} - y_{mean\;exp})^2}$	(9.56)	STATISTICA
q^2	predictive squared correlation coefficients	$= 1 - \dfrac{\sum_{i=1}^{N}(y_{calc} - y_{exp})^2}{\sum_{i=1}^{N}(y_{exp} - y_{mean\;exp})^2}$	(9.57)	ChemProp&R
$q^2_{cv,(N-1)}$	predictive squared correlation coefficients using leave-one-out cross–validation	$= 1 - \dfrac{\sum_{i=1}^{N}(y_{calc}^{(N-1)} - y_{exp})^2}{\sum_{i=1}^{N}(y_{exp} - y_{mean\;exp})^2}$	(9.58)	ChemProp&R
rms_{y^2}	root-mean-square error of correlation	$= \sqrt{\dfrac{\sum_{i=1}^{N}(y_{exp} - y_{calc})^2}{N}}$	(9.59)	STATISTICA
$rms_{cv}^{(N-1)}$	root-mean-square error of prediction using leave-one-out cross validation	$= \sqrt{\dfrac{\sum_{i=1}^{N}(y_{exp} - y_{calc}^{N-1})^2}{N-1}}$	(9.60)	ChemProp & R
bias	systematic error	$= \sum (y_{calc} - y_{exp})/N$	(9.61)	ChemProp
me	mean error	$= \sum abs(y_{calc} - y_{exp})/N$	(9.62)	ChemProp

Chapter 9 - Appendix

mne	maximum negative error	$= \min\sum(y_{calc} - y_{exp})/N$	(9.63) ChemProp
mpe	maximum positive error	$= \max\sum(y_{calc} - y_{exp})/N$	(9.64) ChemProp
F-Test	Fischer significance test	$= \dfrac{\sum_{i=1}^{N}(y_{calc} - \overline{y_{exp}})^2}{\dfrac{\sum_{i=1}^{N}(y_{exp} - y_{calc})^2}{(N-P-1)}}$	(9.65) STATISTICA
T-Test	Statistical significance test	$= \dfrac{\overline{X_1} - \overline{X_2}}{\sqrt{\dfrac{S_1^2}{N_1} + \dfrac{S_2^2}{N_2}}}$	(9.66) OriginPro v.8

$1 \geq r^2 \geq 0$, y_{calc} is predicted value, y_{exp} is experimental value, y_{iR} is recalibrated values, y_{mean} is mean value of experimental values, N is number of data set, $N-1$ is cross validation data set at which $q_{cv}^2 \rightarrow r^2$, P is number of independent property parameters for the proposed prediction model, $N-P$ is the degree of freedom at defined confidence interval, $\overline{X_1}, \overline{X_2}$ are the mean values of both data sets 1 and 2, S_1, S_2 are the variance of both datasets 1 and 2.

The above equations of statistics parameters from (9.56) to (9.66) were calculated from the following references:

Doerffel K. *1966*. Statistik in der analytischen Chemie. VEB Deutscher Verlag für Grundstoffindustrie, Leipzig.

Kessler W. *2007*. Multvariate Dataenanalyse. *WILEY-VCH Verlag GmbH & Co. KGaA*, Weinheim, Germany. pp:89-103.

Moore D S. *2004*. The basic practice of statistics. *Freeman WH Company*, New York.

Puzyn T, Leszczynska D, Leszczynski J. *2009*. Toward the Development of „Nano-QSARs": Advances and Challenges. *Small* 5(22):2494-2509.

Schüürmann G, Ebert R-U, Nendza M, Dearden JC, Paschke A, Kühne R, van Leeuwen K, Vermeire T. *2007*. Prediction fate-relate physicochemical properties. In: *Risk assessment of chemicals: An Introduction*. Springer Science. Dordrecht, Netherland. pp: 375-426.

Schüürmann G, Ebert R-U, Chen J, Wang B, Kühne R. *2008*. External validation and prediction employing the predictive squared correlation coefficient- test set activity mean vs training set activity mean. *J Chem. Inf Model*. 48:2140-2145.

9.22 Applicability Domain and Statistical Performance of Prediction Methods in Literature

MODEL	PHENOLS	PAHs	ORGANOCHLORINES & BIPHENYLS & CHLOROBENZENES	HETEROCYCLIC COMPOUNDS	R-BENZENE (R= H, CH₃, 3(CH₃), CHO), & ANILINES
Schüürmann et al. (2006)	PHENOL	FLUORENE	BIPHENYL	4-VINYLPYRIDINE	PENTABROMOETHYLBENZENE
	4-METHYLPHENOL	NAPHTHALENE	GAMMA-HEXACHLOROCYCLOHEXANE	QUINOLINE	
	4-NONYLPHENOL	ACENAPHTHENE	MIREX	ACRIDINE	
	3-CHLOROPHENOL	FLUORANTHENE	TRANS-1,3-DICHLOROPROPENE	4-AZAPHENANTHRENE	
	2-CHLOROPHENOL	ANTHRACENE	CHLORDANE	BENZO(C)ACRIDINE	
	2,3-DICHLOROPHENOL	9-METHYLANTHRACENE	ALDRIN	2,2'-DIPYRIDYL	
	2,4-DICHLOROPHENOL	BENZO(K)FLUORANTHENE	HEPTACHLOR	PHENAZINE	
	3,5-DICHLOROPHENOL	NAPHTHACENE	p,p'-DDE	2,2'-BIQUINOLINE	
	3,4-DICHLOROPHENOL	PHENANTHRENE	p,p'-DDD	CYROMAZINE	
	2,4,6-TRICHLOROPHENOL	BENZ(A)ANTHRACENE	p,p'-DDT	SIMETONE	
	3,4,5-TRICHLOROPHENOL	7,12-DIMETHYLBENZ(A)ANTHRACENE	2-CHLOROBIPHENYL/PCB 1	ATRATONE	
	2,3,5-TRICHLOROPHENOL	3-METHYLCHOLANTHRENE	2,2'-DICHLOROBIPHENYL/PCB 4	SECBUMETON	
	2,4,5-TRICHLOROPHENOL	1,2,5,6-DIBENZANTHRACENE	2,4'-DICHLOROBIPHENYL/PCB 8	PROMETONE	
	2,3,4,6-TETRACHLOROPHENOL	ACENAPHTHYLENE	2,4,4'-TRICHLOROBIPHENYL/PCB 28	ANILAZINE	
	PENTACHLOROPHENOL	PYRENE	2,2',5-TRICHLOROBIPHENYL/PCB 18	SIMAZINE	
		INDENO(1,2,3-CD)PYRENE	2,2',4-TRICHLOROBIPHENYL/PCB 17	ATRAZINE	
		DIBENZO(A,L)PYRENE	2,3',4-TETRACHLOROBIPHENYL/PCB 70	PROPAZINE	
		BENZO(GHI)PERYLENE	2,2',5,5'-TETRACHLOROBIPHENYL/PCB 52	TERBUTHYLAZINE	
		1-NAPHTYLAMINE	2,2',6,6'-TETRACHLOROBIPHENYL/PCB 54	CYANAZINE	
		2-AMINOANTHRACENE	2,2',4,5'-PENTACHLOROBIPHENYL/PCB 87	TRIETAZINE	
			2,2',4,5,5'-PENTACHLOROBIPHENYL/PCB 101		
			2,2',3,3',5,5'-HEXACHLOROBIPHENYL/PCB 133		
			2,2',3,3',4,4'-HEXACHLOROBIPHENYL/PCB 128		
			2,2',3,4,5,5',6-HEPTACHLOROBIPHENYL/PCB 185		
			METHOXYCHLOR		
			EPCHLOROHYDRIN		
			DIELDRIN		
			ENDRIN		
			HEPTACHLOR EPOXIDE		
			TRIDIPHANE		
			2,3,7,8-TETRACHLORO-DIBENZODIOXINE		
			KEPONE		

Sabljić et al. (1995)	PHENOL	NAPHTHALENE	CHLOROBENZENE	SIMAZINE	BENZENE
	2,3-DICHLOROPHENOL	ANTHRACENE	1,2-DICHLOROBENZENE	PROPAZINE	TOLUENE
	2,4-DICHLOROPHENOL	PHENANTHRENE	1,3-DICHLOROBENZENE	AMETRYN	ETHYLBENZENE
	2,4,6-TRICHLOROPHENOL	TETRACENE	1,4-DICHLOROBENZENE	TERBUTRYN	1,2-DIMETHYLBENZENE
	2,4,5-TRICHLOROPHENOL	PYRENE	1,2,3-TRICHLOROBENZENE	PROMETON	1,3-DIMETHYLBENZENE
	3,4,5-TRICHLOROPHENOL	DIBENZ[1,2,5,6]ANTHRACENE	1,2,4-TRICHLOROBENZENE	ATRAZINE	1,4-DIMETHYLBENZENE
	2,3,4,6-TETRACHLOROPHENOL	FLUORENE	1,3,5-TRICHLOROBENZENE	IPAZINE	PROPYLBENZENE
	PENTACHLOROPHENOL	FLUORANTHENE	1,2,3,5-TETRACHLOROBENZENE	TRIETAZIN	1,3,5-TRIMETHYLBENZENE
	2-CHLOROPHENOL	1-METHYLNAPHTHALENE	PENTACHLOROBENZENE	DIPROPETRYN	1,2,3-TRIMETHYLBENZENE
	3-CHLOROPHENOL	2-METHYLNAPHTHALENE	HEXACHLOROBENZENE	TERBUTHYLAZINE	1,2,4,5-TETRAMETHYLBENZENE
	3,4-DICHLOROPHENOL	1-ETHYLNAPHTHALENE	BIPHENYL	PROMETRYN	N-BUTYLBENZENE
	3,5-DIMETHYLPHENOL	2-ETHYLNAPHTHALENE	2-CHLOROBIPHENYL	CYANAZINE	STYRENE
	2,3,5-TRIMETHYLPHENOL	9-METHYANTHRACENE	2,2'-DICHLOROBIPHENYL	SECBUMETON	CHLOROBENZENE
	4-METHYLPHENOL	BENZ[A]ANTHRACENE	2,4'-DICHLOROBIPHENYL	BIQUINOLINE	1,2-DICHLOROBENZENE
		7,12-DIMETHYLBENZ(A)ANTHRACENE	2,4,4'-TRICHLOROBIPHENYL	PHENAZINE	1,3-DICHLOROBENZENE
		3-METHYLCHOLANTHRENE	2,5,2'-TRICHLOROBIPHENYL		1,4-DICHLOROBENZENE
		2-AMINOANTRACENE	2,4,2'-TRICHLOROBIPHENYL		1,2,3-TRICHLOROBENZENE
		6-AMINOCHRYSENE	2,6,2'-6'-TETRACHLOROBIPHENYL		1,2,4-TRICHLOROBENZENE
		1-AMINONAPHTHALENE	2,5,3',4'-TETRACHLOROBIPHENYL		1,3,5-TRICHLOROBENZENE
			2,5,2',5'-TETRACHLOROBIPHENYL		1,2,3,5-TETRACHLOROBENZENE
			2,3,4,2',5'-PENTACHLOROBIPHENYL		PENTACHLOROBENZENE
			2,4,5,2',5'-PENTACHLOROBIPHENYL		HEXACHLOROBENZENE
			2,3,4,2',3',4'-HEXACHLOROBIPHENYL		BROMOBENZENE
			2,4,5,2',4',5'-HEXACHLOROBIPHENYL		IODOBENZENE
			pp'-DDT		
			pp'-DDE		
			DICHLOROMETHANE		
			TRICHLOROMETHANE		
			TETRACHLOROMETHANE		
			1,1-DICHLOROETHANE		
			1,2-DICHLOROETHANE		
			1,1,1-TRICHLOROETHANE		
			1,1,2-TRICHLOROETHANE		
			1,1,2,2-TETRACHLOROETHANE		
			1,2-DICHLOROPROPANE		
			1,1-DICHLOROETHENE		
			TETRACHLOROETHENE		
			TRICHLOROETHENE		

Tao et al. (1999)

PHENOL	FLUORENE	BIPHENYL	QUINOLINE	BENZENE
2-METHYLPHENOL	NAPHTHALENE	DICHLOROMETHANE	ACRIDINE	TOLUENE
3-METHYLPHENOL	1-METHYLNAPHTHALENE	1,2-DICHLOROETHANE	4-AZAPHENANTHRENE	o-XYLENE
4-METHYLPHENOL	2-METHYLNAPHTHALENE	1,1-DICHLOROETHANE	PHENAZINE	ETHYLBENZENE
3,5-DIMETHYLPHENOL	2-ETHYLNAPHTHALENE	1,2-DICHLOROPROPANE	2,2'-BIQUINOLINE	p-XYLENE
2,3,5-TRIMETHYLPHENOL	1-ETHYLNAPHTHALENE	TRICHLOROMETHANE	HYDROXY ATRAZINE	m-XYLENE
2-CHLOROPHENOL	2,3-DIMETHYLNAPHTHALENE	1,1,2-TRICHLOROETHANE	SECBUMETON	1,2,3-TRIMETHYLBENZENE
3-CHLOROPHENOL	ACENAPHTHENE	1,1,1-TRICHLOROETHANE	PROMETONE	1,3,5-TRIMETHYLBENZENE
4-CHLOROPHENOL	BENZO(A)FLUORENE	TETRACHLOROMETHANE	ANILAZINE	PROPYLBENZENE
2,3-DICHLOROPHENOL	FLUORANTHENE	1,1,1,2-TETRACHLOROETHANE	SIMAZINE	1,2,4-TRIMETHYLBENZENE
2,4-DICHLOROPHENOL	ANTHRACENE	1,1,2,2-TETRACHLOROETHANE	2-CHLORO-4-ISOPROPYLAMINO-6-METHYLAMINO-S-TRIAZINE	1-ETHYL-4-METHYLBENZENE
3,4-DICHLOROPHENOL	9-METHYLANTHRACENE	HEXACHLOROETHANE	ATRAZINE	1,2,4,5-TETRAMETHYLBENZENE
2,4,5-TRICHLOROPHENOL	BENZO(K)FLUORANTHENE	GAMMA-HEXACHLOROCYCLOHEXANE	PROPAZINE	BUTYLBENZENE
2,4,6-TRICHLOROPHENOL	NAPHTHACENE	TOXAPHENE	TERBUTHYLAZINE	1,3,5-TRIETHYLBENZENE
1,2,4,5-TETRACHLOROBENZENE	PHENANTHRENE	TRANS-1,2-DICHLOROETHYLENE	CYANAZINE	STYRENE
3,4,5-TRICHLOROPHENOL	BENZO(B)FLUORANTHENE	1,1-DICHLOROETHYLENE	TRIETAZINE	CHLOROBENZENE
2,3,4,6-TETRACHLOROPHENOL	BENZ(A)ANTHRACENE	TRANS-1,3-DICHLOROPROPENE	IPAZINE	2-CHLOROTOLUENE
PENTACHLOROPHENOL	7,12-DIMETHYLBENZ(A)ANTHRACENE	TRICHLOROETHYLENE		1,4-DICHLOROBENZENE
	3-METHYLCHOLANTHRENE	TETRACHLOROETHYLENE		1,3-DICHLOROBENZENE
	CHRYSENE	CHLORDANE		1,2-DICHLOROBENZENE
	1,2,5,6-DIBENZANTHRACENE	HEXACHLOROCYCLOPENTADIENE		1,3,5-TRICHLOROBENZENE
	1,2,5,6-DIBENZANTHRACENE	ALDRIN		1,2,4-TRICHLOROBENZENE
	PYRENE	HEPTACHLOR		1,2,3,5-TETRACHLOROBENZENE
	INDENO(1,2,3-CD)PYRENE	CHLOROBENZENE		1,2,3,4-TETRACHLOROBENZENE
	BENZ(A)PYRENE	2-CHLOROTOLUENE		PENTACHLOROBENZENE
	PERYLENE	1,4-DICHLOROBENZENE		HEXACHLOROBENZENE

1,3-DICHLOROPROPENE
α-HCH
β-HCH
γ-HCH
ALDRIN
CHLORDANE
MIREX
2,3,7,8-TETRACHLORODIBENZODIOXINE
METHOXYCHLOR
DIELDRIN

Chapter 9 - Appendix

BENZO(E)PYRENE
BENZO(GHI)PERYLENE
1-NAPHTHYLAMINE
2-AMINOANTHRACENE
6-AMINOCHRYSENE

BROMOBENZENE
1,3-DICHLOROBENZENE
1,2-DICHLOROBENZENE
1,3,5-TRICHLOROBENZENE
1,2,4-TRICHLOROBENZENE
1,2,3,5-TETRACHLOROBENZENE
1,2,3,4-TETRACHLOROBENZENE
PENTACHLOROBENZENE
HEXACHLOROBENZENE
p,p'-DDD
p,p'-DDE
p,p'-DDT
2-CHLOROBIPHENYL/PCB 1
3-CHLOROBIPHENYL/PCB 2
2,2'-DICHLOROBIPHENYL/PCB 4
2,4'-DICHLOROBIPHENYL/PCB 8
4,4'-DICHLOROBIPHENYL/PCB 15
2,2',5-TRICHLOROBIPHENYL/PCB 18
2,4,4'-TRICHLOROBIPHENYL/PCB 28
2,2',4-TRICHLOROBIPHENYL/PCB 17
2,2',5,5'-TETRACHLOROBIPHENYL/PCB 52
2,2',6,6'-TETRACHLOROBIPHENYL/PCB 54
2,2',4,5,5'-PENTACHLOROBIPHENYL/PCB 101
2,2',3,4,5'-PENTACHLOROBIPHENYL/PCB 87
2,2',3,4,6-PENTACHLOROBIPHENYL/PCB 88
2,2',3,4,5,5'-HEXACHLOROBIPHENYL/PCB 141
2,2',3,3',5,5'-HEXACHLOROBIPHENYL/PCB 133
2,2',4,4',6,6'-HEXACHLOROBIPHENYL/PCB 155
2,2',3,3',4,4'-HEXACHLOROBIPHENYL/PCB 128
2,2',3,4,4',5,5'-HEPTACHLOROBIPHENYL/PCB 185
2,2',3,3',5,5',6,6'-OCTACHLOROBIPHENYL/PCB 202
METHOXYCHLOR
DIELDRIN

Nguyen et al. (2005)	2,3-DICHLOROPHENOL	NAPHTHALENE	CHLOROBENZENE	ACRIDINE	BENZENE
	2,4-DICHLOROPHENOL	PHENANTHRENE	1,2-DICHLOROBENZENE		TOLUENE
	2,4,6-TRICHLOROPHENOL	ANTHRACENE	1,4-DICHLOROBENZENE		p-XYLENE
	PENTACHLOROPHENOL	FLUORANTHENE	1,3-DICHLOROBENZENE		ETHYLBENZENE
		1-METHYLNAPHTHALENE	1,2,3-TRICHLOROBENZENE		o-XYLENE
		2-METHYLNAPHTHALENE	1,2,4-TRICHLOROBENZENE		1,3,5-TRIMETHYLBENZENE
		1-ETHYLNAPHTHALENE	1,2,3,4-TETRACHLOROBENZENE		1,2,3-TRIMETHYLBENZENE
		2-ETHYLNAPHTHALENE	1,2,4,5-TETRACHLOROBENZENE		1,2,4,5-TETRAMETHYLBENZENE
		9-METHYLANTHRACENE	2-CHLOROBIPHENYL/PCB 1		PROPYLBENZENE
		PYRENE	2,2'-DICHLOROBIPHENYL/PCB 4		BUTYLBENZENE
		NAPHTHACENE	2,4'-DICHLOROBIPHENYL/PCB 8		CHLOROBENZENE
			2,4,4'-TRICHLOROBIPHENYL/PCB 28		1,2-DICHLOROBENZENE
			2,2',5,5'-TETRACHLOROBIPHENYL/PCB 52		1,4-DICHLOROBENZENE
			2,2',3,3',5,5'-HEXACHLOROBIPHENYL/PCB 133		1,3-DICHLOROBENZENE
			2,3',4',5-TETRACHLOROBIPHENYL/PCB 70		1,2,3-TRICHLOROBENZENE
			2,2',6,6'-TETRACHLOROBIPHENYL/PCB 54		1,2,4-TRICHLOROBENZENE
			2,2',4,4',6,6'-HEXACHLOROBIPHENYL/PCB 155		1,2,3,4-TETRACHLOROBENZENE
			2,2',3,5',6-PENTACHLOROBIPHENYL/PCB 95		1,2,4,5-TETRACHLOROBENZENE
			2,2',3',4,5-PENTACHLOROBIPHENYL/PCB 97		
			2,2',4,5,5'-PENTACHLOROBIPHENYL/PCB 101		
			2,2',3,3',4,4'-HEXACHLOROBIPHENYL/PCB 128		
			2,2',3,3',6,6'-HEXACHLOROBIPHENYL/PCB 136		
			2,2',3,3',4,4',5,5'-OCTACHLOROBIPHENYL/PCB 194		
			2,2',3,3',5,5',6,6'-OCTACHLOROBIPHENYL/PCB 202		
			TRICHLOROMETHANE		
			TETRACHLOROMETHANE		
			1,2-DICHLOROETHANE		
			1,2-DIBROMOETHANE		
			1,1,1-TRICHLOROETHANE		
			TRICHLOROETHYLENE		
			1,1,2,2-TETRACHLOROETHANE		
			TETRACHLOROETHYLENE		
			1,2-DICHLOROPROPANE		

Chapter 9 - Appendix

Source	Phenols	PAHs	Chlorinated compounds	N-heterocycles	Benzenes
Poole et al. (1999)	PHENOL 4-METHYLPHENOL 3,5-DIMETHYLPHENOL 2,3,5-TRIMETHYLPHENOL 2-CHLOROPHENOL 3-CHLOROPHENOL 2,3-DICHLOROPHENOL 2,4-DICHLOROPHENOL 3,4-DICHLOROPHENOL 2,4,6-TRICHLOROPHENOL PENTACHLOROPHENOL	NAPHTHALENE 1-METHYLNAPHTHALENE 2-METHYLNAPHTHALENE 1-ETHYLNAPHTHALENE 2-ETHYLNAPHTHALENE ANTHRACENE 9-METHYLANTHRACENE PHENANTHRENE FLUORENE TETRACENE PYRENE BENZ[A]ANTHRACENE 1,2,5,6-DIBENZANTHRACENE BENZO[A]PYRENE	BIPHENYL CHLOROBENZENE 1,2-DICHLOROBENZENE 1,3-DICHLOROBENZENE 1,4-DICHLOROBENZENE 1,2,3-TRICHLOROBENZENE 1,2,4-TRICHLOROBENZENE 1,3,5-TRICHLOROBENZENE 1,2,3,4-TETRACHLOROBENZENE 1,2,3,5-TETRACHLOROBENZENE PENTACHLOROBENZENE HEXACHLOROBENZENE DICHLOROMETHANE TRICHLOROMETHANE TETRACHLOROMETHANE 1,1-DICHLOROETHANE 1,2-DICHLOROETHANE 1,1,1-TRICHLOROETHANE 1,1,2-TRICHLOROETHANE 1,1,2,2-TETRACHLOROETHANE 1,1-DICHLOROETHENE TRICHLOROETHENE TETRACHLOROETHENE 1,2-DICHLOROPROPANE	CARBAZOLE ACRIDINE	BENZENE TOLUENE ETHYLBENZENE 1,2-DIMETHYLBENZENE 1,3-DIMETHYLBENZENE 1,4-DIMETHYLBENZENE n-PROPYLBENZENE 1,3,5-TRIMETHYLBENZENE 1,2,3-TRIMETHYLBENZENE STYRENE 1,2,4,5-TETRAMETHYLBENZENE n-BUTYLBENZENE CHLOROBENZENE 1,2-DICHLOROBENZENE 1,3-DICHLOROBENZENE 1,4-DICHLOROBENZENE 1,2,3-TRICHLOROBENZENE 1,2,4-TRICHLOROBENZENE 1,3,5-TRICHLOROBENZENE 1,2,3,4-TETRACHLOROBENZENE PENTACHLOROBENZENE HEXACHLOROBENZENE BROMOBENZENE IODOBENZENE
KOCWIN Software (log K_{oc})	PHENOL 2-CHLOROPHENOL 3-CHLOROPHENOL 3,4-DICHLOROPHENOL 3,5-DIMETHYLPHENOL	BENZO[A]PYRENE BENZ[A]ANTHRACENE FLUORENE FLUORANTHENE 2-AMINOANTHRACENE 1-METHYLNAPHTHALENE	BIPHENYL 2,2',4-PCB 2,2',5-PCB ALDRIN a-BHC (benzene hexachloride) g-BHC (benzene hexachloride) a-CHLORDANE	CARBAZOLE 7H-DIBENZO[C,G]CARBAZOLE 13H-DIBENZO[A,I]CARBAZOLE QUINOLINE ACRIDINE	STYRENE o-XYLENE n-PROPYLBENZENE IODOBENZENE
Meylan et al. (1992) (information available for validation set only)	2,3,5-TRIMETHYLPHENOL p-CRESOL	1-METHYLNAPHTHALENE 2-ETHYLNAPHTHALENE 1-NAPHTHYLAMINE 6-AMINOCHRYSENE	MIREX 1,3-DICHLOROPROPENE	BENZO[F]QUINOLINE	

- 239 -

Chapter 9 - Appendix

KOCWIN Software (Molecular Topology) Meylan et al. (1992), and (2004)

PHENOL	FLUORENE	BIPHENYL	CARBAZOLE	BENZENE
4-METHYLPHENOL	NAPHTHALENE	DICHLOROMETHANE	1,2,7,8-DIBENZOCARBAZOLE	TOLUENE
3,5-DIMETHYLPHENOL	1-METHYLNAPHTHALENE	1,2-DICHLOROETHANE	7H-DIBENZO[C,G]CARBAZOLE	o-XYLENE
4-NONYLPHENOL	2-METHYLNAPHTHALENE	1,1-DICHLOROETHANE	PYRIDINE	ETHYLBENZENE
2-CHLOROPHENOL	2-ETHYLNAPHTHALENE	1,2-DICHLOROPROPANE	QUINOLINE	p-XYLENE
3-CHLOROPHENOL	1-ETHYLNAPHTHALENE	TRICHLOROMETHANE	ACRIDINE	m-XYLENE
3,4-DICHLOROPHENOL	ACENAPHTHENE	1,1,2-TRICHLOROETHANE	4-AZAPHENANTHRENE	PROPYLBENZENE
2,4-DICHLOROPHENOL	ACENAPHTHYLENE	1,1,1-TRICHLOROETHANE	PHENAZINE	1,3,5-TRIMETHYLBENZENE
2,4,6-TRICHLOROPHENOL	FLUORANTHENE	TETRACHLOROMETHANE	2,2'-BIQUINOLINE	1,2,3-TRIMETHYLBENZENE
2,4,5-TRICHLOROPHENOL	ANTHRACENE	1,1,2,2-TETRACHLOROETHANE	SIMETONE	1,2,4,5-TETRAMETHYLBENZENE
2,3,4,6-TETRACHLOROPHENOL	9-METHYLANTHRACENE	GAMMA-HEXACHLOROCYCLOHEXANE	ATRATONE	BUTYLBENZENE
PENTACHLOROPHENOL	BENZO(K)FLUORANTHENE	ALPHA-HEXACHLOROCYCLOHEXANE	SECBUMETON	STYRENE
2,3,5-TRIMETHYLPHENOL	NAPHTHACENE	1,1-DICHLOROETHYLENE	PROMETONE	CHLOROBENZENE
2,3-DICHLOROPHENOL	PHENANTHRENE	TRICHLOROETHYLENE	ANILAZINE	1,2-DICHLOROBENZENE
3,5-DICHLOROPHENOL	BENZ(A)ANTHRACENE	TETRACHLOROETHYLENE	SIMAZINE	1,4-DICHLOROBENZENE
3,4,5-TRICHLOROPHENOL	7,12-DIMETHYLBENZ(A)ANTHRACENE	CHLORDANE	ATRAZINE	1,3-DICHLOROBENZENE
2,3,5-TRICHLOROPHENOL	3-METHYLCHOLANTHRENE	HEPTACHLOR	PROPAZINE	1,2,3-TRICHLOROBENZENE
2,3,4,5-TETRACHLOROPHENOL	1,2,5,6-DIBENZANTHRACENE	CHLOROBENZENE	TERBUTHYLAZINE	1,3,5-TRICHLOROBENZENE
	PYRENE	1,2-DICHLOROBENZENE	CYANAZINE	1,2,4-TRICHLOROBENZENE
	INDENO[1,2,3-CD]PYRENE	1,4-DICHLOROBENZENE	TRIETAZINE	1,2,4,5-TETRACHLOROBENZENE
	BENZO(A)PYRENE	1,2,3-TRICHLOROBENZENE	IPAZINE	1,2,3,4-TETRACHLOROBENZENE
	1-NAPHTHYLAMINE	1,3,5-TRICHLOROBENZENE		1,2,3,5-TETRACHLOROBENZENE
	2-AMINOANTHRACENE	1,2,4-TRICHLOROBENZENE		PENTACHLOROBENZENE
	6-AMINOCHRYSENE	1,2,4,5-TETRACHLOROBENZENE		HEXACHLOROBENZENE
		HEXACHLOROBENZENE		BROMOBENZENE
		p,p'-DDD		IODOBENZENE
		p,p'-DDE		
		p,p'-DDT		
		MIREX		
		2,2,5,6-TETRACHLORO-1,7-BIS(CHLOROMETHYL)-7-(DICHLOROMETHYL)BICYCLO[2.2.1]HEPTANE		
		1,3-DICHLOROPROPENE		

- 240 -

Chapter 9 - Appendix

5-CHLORO-2-(CHLOROMETHYL)-2-METHYL-3-METHYLENEBICYCLO[2.2.1]HEPTANE
ALDRIN
DIENOCHLOR
1,3-DICHLOROBENZENE
1,2,3,4-TETRACHLOROBENZENE
1,2,3,5-TETRACHLOROBENZENE
PENTACHLOROBENZENE
o,p'-DDE
2-CHLOROBIPHENYL/PCB_1
2,2'-DICHLOROBIPHENYL/PCB_4
2,4'-DICHLOROBIPHENYL/PCB_8
2,4,4'-TRICHLOROBIPHENYL/PCB_28
2,2',5-TRICHLOROBIPHENYL/PCB_18
2,2',4-TRICHLOROBIPHENYL/PCB_17
2,2',6,6'-TETRACHLOROBIPHENYL/PCB_54
2,3',4,5-TETRACHLOROBIPHENYL/PCB_70
2,2',5,5'-TETRACHLOROBIPHENYL/PCB_52
2,2',4,5,5'-PENTACHLOROBIPHENYL/PCB_101
2,2',3,4,5'-PENTACHLOROBIPHENYL/PCB_87
2,2',3,5,6-PENTACHLOROBIPHENYL/PCB_95
2,2',3',4,5-PENTACHLOROBIPHENYL/PCB_97
2,2',4,4',5,5'-HEXACHLOROBIPHENYL/PCB_153
2,2',3,3',4,4'-HEXACHLOROBIPHENYL/PCB_128
2,2',4,4',6,6'-HEXACHLOROBIPHENYL/PCB_155
2,2',3,4,4',5'-HEXACHLOROBIPHENYL/PCB_138
2,2',3,3',5,5'-HEXACHLOROBIPHENYL/PCB_133
2,2',3,4',5',6-HEXACHLOROBIPHENYL/PCB_149
2,2',3,3',6,6'-HEXACHLOROBIPHENYL/PCB_136
2,2',3,4,5,5',6-HEPTACHLOROBIPHENYL/PCB_185
2,2',3,3',5,5',6,6'-OCTACHLOROBIPHENYL/PCB_202
2,2',3,3',4,4',5,5'-OCTACHLOROBIPHENYL/PCB_194
2,2',4,4',6,6'-HEXABROMOBIPHENYL/PBB_155
METHOXYCHLOR

- 241 -

Chapter 9 - Appendix

		EPICHLOROHYDRIN
		ENDRIN
		DIELDRIN
		HEPTACHLOR_EPOXIDE
		TRIDIPHANE
		2,3,7,8-TETRACHLORO-DIBENZODIOXINE
		KEPONE
Bronner et al. (2010a)		
2-CHLOROPHENOL	NAPHTHALENE	CHLOROBENZENE	BENZENE
4-CHLOROPHENOL	1,2-DIMETHYLNAPHTHALENE	1,4-DICHLOROBENZENE	TOLUENE
3,4-DICHLOROPHENOL	ACENAPHTHENE	1,2,4-TRICHLOROBENZENE	ETHYLBENZENE
2,4,5-TRICHLOROPHENOL	FLUORENE	1,2,3,4-TETRACHLOROBENZENE	PROPYLBENZENE
	PHENANTHRENE	1-CHLOROPENTANE	BUTYLBENZENE
		1-CHLOROHEPTANE	PENTYLBENZENE
		1-CHLOROOCTANE	HEXYLBENZENE
Endo et al. (2009a)			
PHENOL	NAPHTHALENE	DICHLOROMETHANE	QUINOLINE	BENZENE
4-ETHYLPHENOL	PHENANTHRENE	TRICHLOROMETHANE		TOLUENE
2,6-DIMETHYLPHENOL	FLUORANTHENE	TETRACHLOROMETHANE		n-PROPYLBENZENE
2,4,6-TRIMETHYLPHENOL	1-NITRONAPHTHALENE	PENTACHLOROETHANE		CHLOROBENZENE
2-CHLOROPHENOL		γ-HEXACHLOROCYCLOHEXANE		1,2-DICHLOROBENZENE
4-CHLOROPHENOL		cis-DICHLOROETHENE		
2,4,5-TRICHLOROPHENOL		TRICHLOROETHENE		
		TETRACHLOROETHENE		
		CHLOROBENZENE		
		1,2-DICHLOROBENZENE		
Franco and Trapp (2008)		
2,4,6-TRICHLOROPHENOL	1-NAPHTHYLAMINE	Atrazine
2,4-DICHLOROPHENOL		Ametryn
		Terbutryn
		Quinoline
		Desmetryn
		PROMETRYN
		PROMETON
		ACRIDINE
		BENZO(F)QUINOLIN

- 242 -

Chapter 9 - Appendix

Neale et al. (2012)
3-CHLOROPHENOL
4-CHLOROPHENOL
p-CRESOL
2,4-DIMETHYLPHENOL
2,4,6-TRIMETHYLPHENOL
4-CHLORO-3-METHYLPHENOL
4-CHLORO-3,5-DIMETHYLPHENOL
NAPHTHALENE
1-METHYLNAPHTHALENE
1-NITRONAPHTHALENE
PYRENE
BENZO[A]PYRENE
DIBENZ[A,H]ANTHRACENE
1-CHLOROHEPTANE
1-CHLOROOCTANE
1-CHLORODECANE
1,2-DICHLOROBENZENE
1,3-DICHLOROBENZENE
1,2,4-TRICHLOROBENZENE
1,4-DICHLOROBENZENE
BIPHENYL
1,2,3,4-TETRACHLOROBENZENE
1,2,4,5-TETRACHLOROBENZENE
PENTACHLOROBENZENE
TETRACHLOROETHENE
DICHLOROMETHANE
BROMOCHLOROMETHANE
DIBROMOMETHANE
BROMODICHLOROMETHANE
DIBROMOCHLOROMETHANE
TETRACHLOROMETHANE
TRIBROMOMETHANE
cis-1,2-DICHLOROETHENE
γ-HCH
HEXACHLOROBENZENE
PCB 18
PCB 28
PCB 52
PCB 77
PCB 101
PCB 118
PCB 138
PCB 153
PCB 180
CARBAZOLE
ATRAZINE
PROPYLBENZENE
1,2-DICHLOROBENZENE
1,3-DICHLOROBENZENE
1,4-DIBROMOBENZENE
1,2,4-TRICHLOROBENZENE
1,4-DICHLOROBENZENE
n-BUTYLBENZENE
n-PENTYLBENZENE
n-HEXYLBENZENE
1,2,3,4-TETRACHLOROBENZENE
1,2,4,5-TETRACHLOROBENZENE
PENTACHLOROBENZENE

Chapter 9 - Appendix

Model	Class	N	r²	q²	rms	F-test	bias	me	mae	mpe
Fragment constant model	PAHs	12	0.92	0.92	0.24	129.57	0.01	0.17	-0.51	0.40
2D Molecular structure	PAHs	12	0.90	0.89	0.28	94.25	-0.06	0.22	-0.55	0.43
KOCWIN Molecular topology	PAHs	12	0.85	0.77	0.41	40.82	0.23	0.31	-0.33	0.73
LSER(Nguyen)exp.	PAHs	9	0.83	0.57	0.54	29.89	0.24	0.40	-0.37	1.04
LSER (Neale 2012)exp.	PAHs	12	0.80	-0.33	0.98	21.59	0.76	0.81	-0.17	1.58
LSER(Nguyen)calc.	PAHs	12	0.76	0.75	0.43	38.45	-0.04	0.34	-0.92	0.50
LSER (Bronner 2010)exp.	PAHs	12	0.75	0.62	0.52	38.53	0.12	0.42	-0.82	0.88
LSER (Poole)exp.	PAHs	10	0.69	0.42	0.61	20.59	0.32	0.45	-0.24	1.36
LSER (Endo 2009 - low Peat)calc.	PAHs	12	0.67	-4.12	1.93	12.49	1.72	1.72	-1.08	0.62
LSER (Poole)calc.	PAHs	12	0.64	0.64	0.51	20.12	-0.05	0.40	-1.08	0.62
LSER (Neale 2012)calc.	PAHs	12	0.64	0.15	0.78	20.40	0.48	0.65	-0.72	1.31
LSER (Endo 2009 - high Peat)calc.	PAHs	12	0.59	0.22	0.75	19.28	0.39	0.62	-0.88	1.21
LSER (Endo 2009 - high Peat)exp.	PAHs	12	0.58	-0.15	0.91	18.09	0.56	0.61	-0.26	2.39
Franco & Trapp model (2008)	PAHs	12	0.50	-0.22	0.94	8.44	-0.68	0.70	-1.79	0.09
KOCWIN Molecular topology	Phenols	11	0.86	0.75	0.42	26.86	-0.25	0.31	-0.98	0.11
LSER (Neale 2012)exp.	Phenols	14	0.70	-1.83	1.39	21.48	0.92	1.18	-0.89	2.33
LSER (Endo 2009 - high Peat)exp.	Phenols	14	0.70	-0.33	0.95	28.52	0.43	0.79	-1.19	1.66
LSER (Endo 2009 - low Peat)exp.	Phenols	14	0.69	-2.81	1.61	17.33	1.35	1.42	-0.50	2.47
LSER (Bronner 2010)exp.	Phenols	14	0.63	-0.21	0.91	28.38	-0.23	0.75	-1.49	1.14
LSER (Poole)exp.	Phenols	12	0.58	0.38	0.58	21.21	-0.20	0.48	-0.93	0.88
LSER(Nguyen)exp.	Phenols	12	0.57	0.15	0.67	23.33	0.05	0.49	-0.88	1.29
LSER(Nguyen)calc.	Phenols	14	0.50	0.12	0.78	24.13	0.00	0.60	-1.11	1.59
Fragment constant model	Phenols	14	0.55	0.38	0.65	19.22	-0.29	0.55	-1.05	0.91
2D Molecular structure	Organochlorines & Biphenyls	11	0.91	0.27	0.71	21.76	-0.63	0.63	-1.03	0.07
LSER (Endo 2009 - high Peat)exp.	Organochlorines & Biphenyls	11	0.89	-1.69	1.36	16.51	1.12	1.14	-0.11	2.30
Franco & Trapp model (2008)	Organochlorines & Biphenyls	11	0.88	0.03	0.82	15.39	-0.73	0.73	-1.29	0.35
LSER (Endo 2009 - low Peat)calc.	Organochlorines & Biphenyls	11	0.87	-15.50	3.37	10.58	3.07	3.07	1.12	4.83
Fragment constant model	Organochlorines & Biphenyls	11	0.86	0.73	0.43	48.93	-0.21	0.34	-0.79	0.47
LSER (Poole)calc.	Organochlorines & Biphenyls	11	0.85	0.67	0.48	46.01	0.17	0.38	-0.50	0.94
LSER (Endo 2009 - low Peat)exp.	Organochlorines & Biphenyls	11	0.84	-6.43	2.26	12.85	1.84	1.84	0.13	4.41

Chapter 9 - Appendix

Model	Compound class	n								
LSER (Bronner 2010)/exp.	Organochlorines & Biphenyls	11	0.83	0.25	0.72	28.38	0.44	0.58	-0.68	1.21
LSER (Endo 2009 - high Pearl)/calc.	Organochlorines & Biphenyls	11	0.82	-3.41	1.74	13.69	1.45	1.45	0.12	3.00
LSER (Neale 2012)/exp.	Organochlorines & Biphenyls	11	0.78	-4.46	1.94	12.33	1.69	1.69	0.07	2.76
LSER(Nguyen)calc.	Organochlorines & Biphenyls	11	0.77	0.50	0.59	33.51	0.20	0.46	-0.95	0.92
LSER (Neale 2012)/calc.	Organochlorines & Biphenyls	11	0.76	-4.14	1.88	12.92	1.57	1.58	-0.02	3.06
LSER (Bronner 2010)/calc.	Organochlorines & Biphenyls	11	0.75	-0.20	0.91	22.90	0.46	0.74	-1.15	1.80
Molecular connectivity indices	Organochlorines & Biphenyls	11	0.75	0.65	0.49	16.04	-0.22	0.35	-1.23	0.21
Fragment constant model	Heterocyclic compounds	12	0.69	-0.38	0.63	14.76	-0.52	0.52	-1.17	0.13
LSER (Endo 2009 - low Pearl)/calc.	Heterocyclic compounds	12	0.66	-0.67	0.69	21.97	0.25	0.58	-1.39	0.94
LSER (Poole)/calc.	Heterocyclic compounds	12	0.66	-1.41	0.83	12.55	-0.73	0.73	-1.43	-
LSER (Neale 2012)/calc.	Heterocyclic compounds	12	0.63	-4.06	1.20	12.43	-1.07	1.07	-1.91	0.26
LSER (Endo 2009 - high Pearl)/calc.	Heterocyclic compounds	12	0.62	-6.69	1.48	11.52	-1.35	1.35	-2.09	0.34
2D Molecular structure	Heterocyclic compounds	12	0.61	0.02	0.53	18.89	-0.33	0.34	-1.46	0.04
LSER (Poole)/exp.	Heterocyclic compounds	10	0.59	-0.12	0.61	15.87	-0.32	0.38	-1.61	0.18
LSER(Nguyen)calc.	Heterocyclic compounds	12	0.59	-2.25	0.96	12.23	-0.84	0.84	-1.65	0.42
LSER(Nguyen)exp.	Heterocyclic compounds	10	0.51	-0.97	0.81	14.92	-0.15	0.56	-1.97	0.70
LSER (Endo 2009 - low Pearl)/exp.	Heterocyclic compounds	5	0.86	-0.39	1.18	6.88	0.85	0.91	-0.15	1.61
LSER (Poole)/exp.	Halobenzenes	5	0.84	0.70	0.54	10.49	-0.33	0.40	-0.90	0.18
Fragment constant model	Halobenzenes	5	0.84	0.66	0.58	8.40	-0.37	0.41	-1.00	0.09
LSER (Endo 2009 - high Pearl)/exp.	Halobenzenes	5	0.84	0.67	0.57	15.93	0.08	0.49	-0.62	0.65
KOCWIN Molecular topology	Halobenzenes	5	0.84	0.42	0.76	3.04	-0.48	0.54	-1.14	0.10
Molecular connectivity indices	Halobenzenes	5	0.84	0.04	0.98	2.27	-0.66	0.72	-1.39	0.15
2D Molecular structure	Halobenzenes	5	0.83	0.75	0.50	10.56	-0.25	0.36	-0.86	0.28
LSER (Neale 2012)/exp.	Halobenzenes	5	0.83	0.58	0.65	13.68	0.16	0.56	-0.62	0.78
LSER (Bronner 2010)/exp.	Halobenzenes	5	0.82	0.68	0.56	13.67	-0.23	0.42	-0.79	0.28
LSER(Nguyen)exp.	Halobenzenes	5	0.82	0.57	0.65	9.53	-0.42	0.45	-1.03	0.08
Franco & Trapp model (2008)	Halobenzenes	5	0.81	0.70	0.55	5.81	-0.24	0.39	-0.91	0.35
KOCWIN (log K_{ow})	Halobenzenes	5	0.80	0.69	0.56	8.46	-0.29	0.44	-0.89	0.37
Fragment constant model	Benzene & R-benzenes	3	1.00	0.92	0.24	6.98	-0.10	0.17	-0.31	0.10
LSER (Endo 2009 - low Pearl)/exp.	Benzene & R-benzenes	4	0.78	0.76	0.34	7.12	-0.09	0.19	-0.55	0.19
LSER (Endo 2009 - low Pearl)/calc.	Benzene & R-benzenes	4	0.55	0.38	0.54	4.34	-0.04	0.38	-0.84	0.29
LSER (Poole)/exp.	Benzene & R-benzenes	4	0.53	-0.12	0.73	1.76	-0.47	0.53	-0.98	0.11

Chapter 9 - Appendix

Model	Class	N	r²	q²	rms	bias	me	mne	mpe	F-test
LSER (Endo 2009 - low Peat)calc.	Anilines	3	0.69	-1.53	1.08	2.06	-0.28	0.88	-0.95	0.89
LSER (Endo 2009 - low Peat)exp.	Anilines	3	0.68	-4.98	1.66	1.72	0.03	1.26	-1.03	1.94
Franco & Trapp model (2008)	Anilines	3	0.67	-2.73	1.31	1.30	-0.98	0.98	-1.51	0.49
LSER (Poole)exp.	Anilines	3	0.67	-3.02	1.36	1.46	-0.93	0.93	-1.54	-
LSER (Endo 2009 - high Peat)exp.	Anilines	3	0.67	-6.56	1.86	1.48	-0.98	1.42	-1.81	0.11
LSER (Endo 2009 - high Peat)calc.	Anilines	3	0.66	-5.01	1.66	1.42	-1.08	1.12	-1.78	0.67
LSER (Neale 2012)calc.	Anilines	3	0.66	-6.05	1.80	1.33	-1.26	1.26	-1.94	0.06
LSER (Neale 2012)exp.	Anilines	3	0.65	-5.77	1.76	1.51	-0.87	1.37	-1.72	-
LSER (Bronner 2010)exp.	Anilines	3	0.65	-8.49	2.09	1.29	-1.44	1.44	-2.19	0.21
Molecular connectivity indices	Anilines	3	0.64	-3.31	1.41	1.31	-1.04	1.04	-1.63	0.75
2D Molecular structure	Anilines	3	0.63	-0.83	0.92	1.43	-0.64	0.64	-1.17	0.15
LSER(Nguyen)calc.	Anilines	3	0.63	-3.41	1.42	1.32	-1.04	1.04	-1.65	0.42
LSER (Poole)calc.	Anilines	3	0.62	-3.48	1.43	1.26	-1.07	1.07	-1.66	0.28
LSER (Bronner 2010)calc.	Anilines	3	0.62	-9.42	2.19	1.18	-1.66	1.66	-2.35	0.37
LSER(Nguyen)exp.	Anilines	3	0.60	-3.07	1.37	1.37	-0.96	0.96	-1.62	0.52
KOCWIN (log K_{oc})	Anilines	3	0.58	-3.57	1.45	1.44	-0.91	0.98	-1.69	0.76

N is no of points, r^2 is squared correlation coefficient, q^2 is predictive squared correlation coefficient, rms is root-mean-square error of correlation, bias is systematic error, me is mean error, mne is maximum negative error, mpe is maximum positive error, F-test was calculated at $\alpha = 0.05$.

Bronner G, Goss KU. ***2010a***. Prediction sorption of pesticides and other multifunctional organic chemicals to soil organic carbon. *Environ. Sci. Technol.*45: 1313-1319.

Endo S, Grathwohl P, Haderlein SB, Schmidt TC. ***2009a***. LFERs for soil organic carbon-water distribution Coefficient (K_{oc}) at environmentally relevant sorbate concentrations. *Environ. Sci. Technol.* 43: 3094-3100.

Franco A, Trapp S. ***2008***. Estimation of the soil-water partition coefficient normalized to organic carbon for ionisable organic chemicals. *Environmental Toxicology and Chemistry* 27(10): 1995-2004.

Meylan WM. ***2004***. KOWWIN 1.67. *Syracuse Research Corporation*, Syracuse, NY.

Meylan WM. *EPSUITE*. *Syracuse Research Corporation*, Syracuse, NY. http://www.syrres.com/esc/est_soft.htm

Meylan WM, Howard PH, Boethling RS.***1992***. Molecular topology/fragment contribution method for prediction soil sorption coefficients. *Environ. Sci. Technol.* 26:1560-1567.

Neale PA, Escher BI, Goss KU, Endo S. ***2012***. Evaluating dissolved organic carbon-water partitioning using polyparameter linear free energy relationships: Implications for the fate of disinfection by-products. *Water Research*. 46: 3637-3645.

Nguyen TH, Goss KU, Ball WL. ***2005***. Poly parameter linear free energy relationships for estimating the equilibrium partition of organic compounds between water and the neutral organic matter in soils and sediments. *Critical Review. Environ. Sci. Technol*. 39(4): 913-924.

Poole SK, Poole CF. ***1999***. Chromatographic models for the sorption of neutral organic compounds by soil from water and air. *J. of Chromatography A* 845: 381-400.

Sabljić A, Güsten H, Verhaa H, Hermens J. ***1995***. QSAR modelling of soil sorption. Improvements and systematic of log k_{oc} vs. log k_{ow} correlations. *Chemosphere* 31: 4489-4514.

Schüürmann G, Ebert RU, Kühne R. ***2006***. Prediction of the sorption of organic compounds into soil from molecular structure. *Environ. Sci. Technol*. 40: 7005-7011.

Tao S, Piao H, Dawson R, Lu X, Hu H. ***1999***. Estimation of organic carbon normalized sorption coefficient (K_{oc}) for soils using the fragment constant method. *Environ. Sci. Technol*.33: 2719-2725.

Rehab Mansour

Wissenschaftliche Veröffentlichungen

- Poster: Der Einfluss des pH-Wertes auf das Sorptionsverhaltens von Xenobiotika SETAC/GDCH (Neue Problemstoffe in der Umwelt), Dritte Gemeinsame Jahrestagung. Goethe- Universität Frankfurt, Frankfurt am Main, 23. - 26. September 2008. (Veröffentlicht)

- Poster: Sorption of Xenobiotics to Humic Acids and its Dependence on pH and pK_a, SETAC Europe 19th Annual Meeting. Göteborg, Schweden, 31. Mai – 3. Juni 2009. (Veröffentlicht)

- Präsentation: Introduction to Soil Sorption of Xenobiotics to Humic Acids and its Impact on Photodegradation in Neutral Water. Department Ökologische Chemie des Helmholtz-Zentrums für Umweltforschung (UFZ), Leipzig, 28. Mai 2007.

- Präsentation: Soil Sorption of Xenobiotics and its Impact on pH. Department Ökologische Chemie des Helmholtz-Zentrums für Umweltforschung (UFZ), Leipzig, 20. Oktober 2008.

I want morebooks!

Buy your books fast and straightforward online - at one of the world's fastest growing online book stores! Environmentally sound due to Print-on-Demand technologies.

Buy your books online at
www.get-morebooks.com

Kaufen Sie Ihre Bücher schnell und unkompliziert online – auf einer der am schnellsten wachsenden Buchhandelsplattformen weltweit! Dank Print-On-Demand umwelt- und ressourcenschonend produziert.

Bücher schneller online kaufen
www.morebooks.de

OmniScriptum Marketing DEU GmbH
Heinrich-Böcking-Str. 6-8
D - 66121 Saarbrücken

Telefax: +49 681 93 81 567-9

info@omniscriptum.de
www.omniscriptum.com

Printed by Books on Demand GmbH, Norderstedt / Germany